RAIL VEHICLE TESTING

Bringing to Life My Experiences Of Railway Vehicle Testing During The 1980s & 1990s

DAVE BOWER

Salient Books

British Library Cataloguing In Publication Data

A Record of this Publication is available from the British Library

ISBN 978-1-9999356-0-3

First Published 2018 by

Salient Books,
21, Collington Street, Beeston, Nottingham, NG9 1FJ

www.salientbooks.co.uk

Cover Photo:- Test Car 1 (ADW150375) during a test run near
Syston Junction, 20th June 1985.

Cover Design © D.Bower and H.Osborne 2018

ACKNOWLEDGEMENTS

The first acknowledgement should be to my mother, Clare, who instilled into me from a very early age to write everything down and not to throw anything away. Although I did not think so at the time, when I was writing in my diaries and subsequently keeping them safe with my photograph collection in boxes in the loft, this has provided the source for the majority of the material for this book. Further acknowledgements need to go to a dear friend Chris O'Donnell who played a huge part in giving me the start in a career in testing, sharing with me his enthusiasm for everything to do with rail vehicle testing when I first joined the Testing Section during my apprenticeship in 1983. Also to Ruth for supporting me with my interest in railways, and dissuading me from throwing things out. Supporting information for the book has come from Mike Fraser who also kept records and took photographs during the series of Foster Yeoman wagon tests in 1989 and 1990 whilst working for the wagon manufacturer O&K; from Hugh Searle and his excellent photographic account of the Foster Yeoman Megatrain trial in 1991, and from Richard Hobson at Serco Rail Technical Services in Derby for his assistance in access to the testing archives which has helped me confirm a few test details which had missed my diary logs.

All photographs and diagrams in this book are from the author's collection or reproduced as credited with permission courtesy of Mike Fraser, Hugh Searle and Serco Rail Technical Services.

INTRODUCTION

My interest in railways started at a very early age in the 1960s; I was born in the upstairs flat at Holmefield which backs onto the A6 in Darley Dale, Derbyshire. The front of the house 'Holmefield' which is now converted into a guest house, and indeed the room in which I was born, overlooks the Derwent Valley and the Matlock to Buxton railway line between Darley Dale and Rowsley, facing towards Stanton-in-Peak.

Fig.1 - The view from Holmefield, early 1963

By the age of one, I was apparently very happy to be sat on my father's knee looking out of the upstairs flat window watching the Blue Pullman trains speed by and freights trundling along at the bottom of the garden. The photo above of an ex LMS 8F locomotive, running light engine south towards Darley Dale, was recorded by my late mother Clare Osborne (1938-2015), photographed from the upstairs flat at Holmefield, looking out from the living room window west towards Stanton-in-Peak, on a winter afternoon in early 1963. The interest in railways was also helped by my grandfather Stanley Beaver, who became Professor of Geography at Keele University in Staffordshire, but also had a keen interest in railways and railway photography. When in his 30s he had become a member and regular attendee of the Stephenson Locomotive Society (SLS), and supported many events, trips and tours during the 1930s across the UK and into Europe. His interest in locomotives of any origin or type was clear from his vast collection of photographs he took during the SLS trips, particularly those photos of the more unusual types of locomotives, such as Pannier Tank No.1729, a GWR 1854 Class of Dean design built in 1892; and an LMS Diesel 0-6-0 No.1831, he recorded at Derby Shed (17A) on 1st September 1935. Stanley was elected as General Secretary of the SLS in January 1937, arranging tours and depot visits to locations such as Crewe South Shed, Swansea Victoria, Swindon Shed, Eastleigh Works, Bromley South and a four day tour of Scotland in June 1937 including visits to Kyle of Lochalsh, Mallaig, Cowlairs, Alloa, Dunfirmline, Thornton and St Margarets. He remained in the position as General Secretary until August 1939, however continued to support and join SLS tours and events until the early 1980s.

Fig.2 - Stanley Beaver on the footplate of Pannier Tank No.1729, July 1934

During 1965 we moved away from Darley Dale to where I grew up in West Bridgford south of Nottingham, and although away from the daily sight of the railway my interest remained. I have happy memories through my childhood at my grandparents' house perusing the vast collection of railway magazines, diaries, books and photos that my grandfather had accumulated over the years.

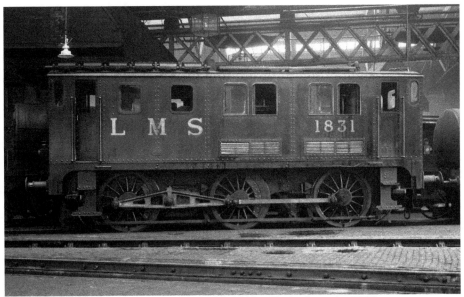

Fig.3 - LMS Diesel 0-6-0 No.1831 at Derby, September 1935

On leaving school at 16, I applied for a number of apprenticeships and was lucky enough to be selected to join the British Railways Board (BRB), Director Mechanical & Electrical Engineering (DM&EE) department 5 year technicians apprenticeship scheme; starting my career on the railways at the Railway Technical Centre (RTC) in Derby in October 1980. The apprenticeship scheme included on and off-the-job training, starting at the Normanton Road annex of the Derby College of Further Education with practical off-the-job training covering basic engineering practices, machining, electrical theory, materials and tooling selection. The practical training was intermixed with theory study learning at the Wilmorton, and Kedleston Road college sites, successfully completing an HNC (Higher National Certificate) in Engineering Design in 1983. In the early 1980s the Railway Engineering School on London Road in Derby provided railway vehicle comprehensive training courses covering the theory of mechanical train systems, brake systems, electrical systems and equipment used on Traction and Rolling stock. I attended two of these residential courses which provided a good basis for the next stage of my apprenticeship in practical 'on-the-job' training. I spent between four and five month periods on-the-job at each of three main British Rail Engineering Limited (BREL) workshops in order to gain hands-on experience on all types of traction and rolling stock. This phase of my apprenticeship started in early 1982 at Swindon (BREL) Works, where I assisted with work on locomotives and multiple units; including experience working on the heavy overhaul of Class 08 shunting locomotives, various types of 1960s built Diesel Multiple Units (DMUs) and also the Class 411/5 (4-CEP) Electric Multiple Unit (EMU) refurbishment programme. Moving then to Crewe (BREL) works my experience working on larger main-line locomotive heavy overhaul and repair continued on Class 37, Class 40, Class 47 and Class 85 locomotives. The last of the three periods of workshop experience during my apprenticeship was at Tinsley Traction depot (between Sheffield and Rotherham), where running and heavy maintenance activity was carried out on a large variety of locomotives such as Class 20, Class 31, Class 45 and the then relatively new Class 56 type 5 heavy freight locomotives.

My first foray into the world of rail vehicle testing was back at Derby in February 1983, when I had the opportunity as part of a placement during my apprenticeship, to spend about six months with the DM&EE Testing Section. The team I joined was predominantly responsible for locomotive testing and at the time the first new Class 58 locomotive (58001) had arrived at the RTC Derby for Type Acceptance testing. With the help of my good friend Chris O'Donnell, I learnt a huge amount about the interesting, and at the time very exciting depth of rail vehicle testing that was being undertaken by the section. I spent time learning the skills of instrumentation selection and fitting, including strain gauging, and undertaking static and on-track tests as part of the introduction of the new heavy freight locomotives. Some more details of the Class 58 locomotive tests that I was involved in are included in later chapter of this book.

In September 1983 I finished my allotted period with the Testing Section and returned to the less hands-on work within the Freight Design team in Trent House at the Railway Technical Centre. Working for the BRB Freight Engineer, David Russell, I supported the interface between the DM&EE design team and Crewe (BREL) Works engineers during the preparation and conversion of the first Class 37 locomotive to be fitted with Electric Train heating (ETH). Locomotive number 37268 was selected for the first conversion, which was one of the later batch of Class 37s, built in the English Electric Vulcan Foundry - Works Number EE/VF3528/D957 and entered service in February 1965 numbered as D6968. The whole fleet of Class 37 locomotives were built over quite a few years at two different factories and by 1983 many had completed over twenty years in service. Design differences were applied as the production progressed and subsequent visits to works for modifications and accident repairs attention, meant that there were many detail differences between the actual configuration of many of the locomotives and the drawing records available. The differences were not only limited to the main visual variations (i.e. the nose end doors, headcode boxes and warning horn positions), but also internal developments, steam heat boiler removals, air brake system installation, etc. These modifications were not necessarily applied consistently to individual locomotives, therefore resulting in variations in pipe and electrical cable conduit routing. As the work commenced on 37268, it was clear that the design being finalised in the DM&EE drawing offices in Derby in preparation for the Electric Train heating (ETH) conversion, needed some adjustment to align with the current state of the locomotive. The freight design team sent me to work with the DM&EE photographer, making regular weekly trips across to Crewe to assess, record and photograph the differences found as the preparations for the conversion gathered pace, such that the necessary changes could be made to the conversion drawings. In the end, the conversion work on 37268 which also included fitting of extra capacity twin fuel tanks, was not completed until May 1985. The locomotive was renumbered 37401 and transferred to the Crewe works test section and completing dynamic tests runs to Llandudno Junction during mid 1985; although I was not directly involved with the final testing of 37401, as by that time I was back in Derby supporting another memorable project that I worked on during my time with the DM&EE Freight Design section. This was to support a new locomotive project design and development in which British Rail worked closely with and supported General Motors Electro Motive Division (EMD) in the development their JT26CW-SS design locomotive for use in the UK, which became the Class 59s built for Foster Yeoman. Working with the lead engineers in the DM&EE design team at Derby, I supported gathering evidence for, and the validation of the structural underframe loading calculations for the proposed new design. My final project work before completing my apprenticeship and securing a post as a Technical Officer (TO) back on the DM&EE Testing Section, was to build a 12th scale replica of the new Class 59 locomotive cab to support finalisation of the shape of the exterior design. This was constructed in balsa wood and assisted the design engineers and Foster Yeoman in visualising how the new locomotives would look before the drawings were finalised.

Joining the DM&EE Testing Section at the end of my apprenticeship opened up a new chapter in my career. This book takes a look back through the wide range of traction and rolling stock testing experiences I was lucky enough to encounter during the 1980s and 1990s, within the hugely varied and interesting field of rail vehicle testing. We will take a look at the test coaches and locomotives used to facilitate the on-track testing and the reasons why testing was undertaken, investigating the static and dynamic tests associated with freight vehicle running characteristics, braking systems and the structural integrity of freight vehicles. There is also an insight into on-track plant vehicle testing, passenger rolling stock and locomotive tests. The later chapters take a look at investigations and modifications assessments as well as some special tests and events that I undertook. Examples are provided within each subject from my experiences during the tests in which I was involved.

Fig.4 - The author preparing an Arbel Car Carrier for Testing, May 1993 [Serco]

Dave Bower, CEng MIMechE
March 2017

CONTENTS

LIST OF TABLES

LIST OF PHOTOGRAPHS AND DIAGRAMS

TEST TRAIN LOCOMOTIVES

RUNNING CHARACTERISTICS

BRAKING SYSTEM TESTS

RAIL VEHICLE STRUCTURAL ASSESSMENT

ON-TRACK PLANT VEHICLE TESTING

TESTING OF PASSENGER ROLLING STOCK

APPENDIX 1 - SUMMARY OF TESTING CAREER

ACCEPTANCE TESTING OF NEW ROLLING STOCK

The DM&EE department within British Rail occupied many of the buildings at the south end of the Railway Technical Centre site in the early 1980s, including the Derwent House, Trent House and the Engineering Development Unit (EDU) workshops. The Testing Section along with other specialist DM&EE functions used the EDU workshops to support activities such as the trial modification of, and acceptance testing of many types of traction and rolling stock. The type acceptance of new or modified traction and rolling stock in the 1980s involved scrutiny of the design to ensure compliance with the DM&EE standards applicable at the time. Often this involved gathering information by testing to validate calculations or performance either statically within the EDU workshop, or during dynamic on-track operations on test lines or the main railway network. The EDU workshop provided very versatile facilities including deep long pits with lighting for under-vehicle access, overhead cranes, vehicle lifting and jacking facilities, the bogie rotation test rig and the vehicle weighbridge. The workshop had four roads plus the weighbridge, and was long enough to accommodate four Mk1 coach lengths. The picture below in Fig.5 shows a variety of vehicles undergoing modification works and testing in 1987 including MkIII coach 12140 being fitted with the prototype SIG bogies and an HST power car DVT conversion; the first Class 155 Sprinter DMU had arrived for type acceptance testing as had the new 80 tonne container flat wagon from Finland with GPS20 bogies.

Fig.5 - EDU Workshop at the RTC Derby, 8th Aug 1987

Prior to embarking on any on-track testing of new or modified rolling stock, it was essential to establish that the vehicle and physical characteristics were compliant with any static test criteria. The table below provides examples of the types of static tests that were normally undertaken on new or modified traction and rolling stock.

Test (Static)	Types of Vehicle	Description of Test
Weighing	All vehicle types	Measuring the static vertical wheel load at each wheel whilst on a straight and level track section.
Torsional stiffness	All vehicle types	The measurement of the torsional stiffness of the vehicle suspension / body to determine the ability to traverse a defined track twist and dip rail joint without causing greater than 60% off-loading of the nominal vertical wheel load at any wheel.
Bogie rotation	All types of bogie vehicles, including wagons, on-track plant, coaches, locomotives and multi-bogie wagons	The rotation of the bogies must be flexible enough to allow negotiation of curves however sufficient rotational torque must be maintained to control stability at speed. Clearance between the bogies, any bogie mounted equipment, and the underside of the vehicle could be checked during the test, at all angles of rotation of the bogie, to simulate traversing the designed minimum radius curvature.
Brake block force	Block brakes vehicle types	The clamping force between the brake block and the wheel tread was measured to ensure compliance with design values.
Brake pad force	Disc brake fitted vehicles	The clamping force between the brake pad and the brake disc was measured to ensure compliance with design values.
Handbrake / parking brake	All vehicle types fitted with a handbrake / parking brake	The measurement of the effectiveness of a hand brake / parking brake to hold the vehicle on a specified gradient without the vehicle rolling away.
Brake system application and release timings	Air brake fitted vehicles	Checking that the length of time it takes for the brakes to apply and release of each vehicle was within specified limits
Body sway / pantograph sway testing	Usually undertaken on vehicles fitted with secondary suspension.	To verify the kinematic gauge envelope of a vehicle; this is the outline of the space occupied by a rail vehicle when in motion, including the effects of suspension movements, body roll, body sway, track cant, etc.
Wedge tests	Any type of vehicle	A wedge test could be used as part of investigations into the dynamic behaviour of a vehicle or in order to establish suspension natural frequency and critical damping levels.

Test (Static)	Types of Vehicle	Description of Test
Wheel tread profile measurements	All vehicle types	The shape of the wheel tread profile is very important to control the way the wheelset runs along the rail. Records were taken when vehicles were presented for test when new to ensure the profile shape matches the design, or when vehicles were tested with worn profiles. If a vehicle exhibited instability during a ride test, measurements to inform investigation into rectification.

Table 1 - Table of typical Static Tests

Once all the relevant static tests had been assessed to be compliant, the on-track testing could begin. The table below highlights the types of dynamic on-track testing that I carried out during my time with the testing section, mostly on freight and on-track plant vehicles but also including tests on some locomotives and passenger rolling stock.

Test (On-Track)	Types of Vehicle	Description of Test
Resistance to derailment	All vehicle types	Ensuring that the forces between the wheels and the rails are maintained within acceptable level so as not to derail. Establishing the safe running stability at the maximum designed operating speed.
Braking Systems Performance	All vehicle types	Brake stopping distance performance or wheel slide protection system functionality. Ensuring all traction and rolling stock stopped in accordance with the signal spacing distances on the railway infrastructure. Also ensuring that brake performance of vehicles was not too severe as to cause wheel tread damage; and where fitted, wheel slide protection systems perform in a manner to protect from wheel tread damage.
Structural integrity	Any type of vehicle.	Supporting the validation of the design, or fatigue life prediction of the vehicle structure or components tested.
Pantograph current collection performance	Electric multiple units of electric locomotives	Measurements of the ability of a pantograph to maintain current collection from the contact wire; i.e. to maintain sufficient uplift force against the contact wire to minimise arcing whilst not exceeding the maximum permissible value when running up to maximum speed.
Ride comfort	All passenger carrying rolling stock.	Passenger saloon area vibration comfort level measurement.

Test (On-Track)	Types of Vehicle	Description of Test
Traincrew / Operator environment	Driving cabs of locomotives, on-track plant and multiple units. Other traincrew areas in rolling stock such as kitchen areas.	Vibration and noise level measurements in traincrew areas, including seated and standing, to ensure the levels to which staff are exposed during a working day are maintained within limits.
EE&CS	All traction and rolling stock with electrical traction or associated equipment.	Assessment of electro-magnetic compatibility (EMC) and radio frequency interference (RFI) emissions whilst a vehicle or multiple vehicles are operating in all representative states of operation, including degraded modes of operation.
Traction performance	Locomotives, multiple units, self-propelled on-track plant	Acceleration performance, in different modes of operation including degraded modes and with tare and laden train loads
Engine & Auxiliary Systems	Multiple units and locomotives	Tests such as compressor duty cycles, engine cold and hot starting, shut-down and stall testing

Table 2 - Table of typical Dynamic On-Track Tests

The majority of tests, particularly for freight rolling stock were conducted with the vehicle under test in both the Tare condition (i.e. no payload) and also in the Fully Laden or Crush Laden condition. Loading of the vehicles for testing purposes could be carried out in a number of ways, all of which had their difficulties and benefits. For example sandbags were easily handled when loading passenger vehicles, however the spillage of sand from a bag inside a passenger vehicle was less than ideal, because sand could get into places from where it was not easily removed afterwards. In the case of passenger vehicles therefore ballast bags (such as pea gravel) were preferred. Here are a few examples of methods used for the loading of other types of vehicle.

Types of Vehicle	Loading Method
Mineral Wagons	Ballast, sand, steel weights
Tank Wagons	Water
Container Wagons	Containers filled with steel weights
Car Carrier Wagons	Sandbags or steel weights
Passenger vehicles	Sand or ballast bags, steel weights or water capsules

Table 3 - Typical methods of loading of vehicles for testing

When passenger vehicles were tested, it was normal practice to assess the performance in the tare condition, fully seated condition and also in the crush loaded condition. Any water tanks were filled and where fitted, the Controlled-Emission Toilets (CET) tanks remained empty. Most dynamic tests involved collecting data in one form or another during the tests; this was normally carried out by installing test equipment onto the vehicle under test. The data measurement, recording and analysis equipment was installed within the vehicle under test where possible; or in the case of freight vehicles or on-track plant vehicles, the equipment was located within a separate test coach. The design of the vehicle type, the test train formation, the vehicle load condition and the scope of the tests to be undertaken, defined what combination of tests could be undertaken at the same time. For example an on-track machine such as a Tamper, which was self-powered with a maximum designed operating speed of 45 mile/h underwent a single dynamic test run arranged to incorporate resistance to derailment, brake stopping distance performance and traincrew / operator environment assessment. On the other hand, a freight vehicle with a payload capacity of 80 tonnes had at least four separate tests to cover resistance to derailment and braking performance (by the slip/brake method), in both tare (unloaded) and fully laden load conditions.

The following chapters provide a more detailed insight into examples of the static and dynamic testing that I lead, or played a key role in during my time with on the Vehicle Testing Section. We will take a look at the test coaches and locomotives used to support the operation of test trains; then at freight vehicle testing, which for many years formed the main part of my work, in particular undertaking running characteristics, brake system and structural assessment testing. Intermixed with the freight vehicle tests were various on-track plant vehicle tests, passenger stock and locomotive tests for which there are separate chapters. One of the biggest projects that I was involved in was that of the development and acceptance of the Mk IV coaches for the East Coast Main line, the testing activities for which are described in the passenger rolling stock chapter. Further chapters cover examples of more unusual vehicle types such as road & rail freight trains, special purpose vehicles, tests in the Channel Tunnel and special test runs such as the Foster Yeoman 12000 tonne Megatrain trial. For reference there is an appendices containing an account from my diaries logged between 1983 and 1999 providing some more history of dynamic tests that I was on board whilst undertaking tests as an Engineer or as an Officer in Charge of the tests.

TEST COACHES

The British Rail Director of Mechanical & Electrical Engineering (DM&EE) test car fleet during the 1980s and 1990s consisted of a number of converted passenger coaches adapted for various different types of on-track testing. For the majority of the time I spent at Derby with the Test Section my work involved undertaking static and dynamic on-track testing predominantly using Test Cars 1, 2, 3, 6 and 10. All five of these coaches were adapted to support various types of testing and therefore had very different interior layouts.

Test Car	Purpose
Test Car 1	Predominantly used for freight and on-track plant vehicle dynamic ride and vehicle behaviour testing until 1997.
Test Car 2	Adapted specifically for slip/brake testing, but also used for various other traction and rolling stock tests.
Test Car 3	More commonly known as 'Mentor' which stands for Mobile Electrical Network Testing, Observation and Recording, with flat roof areas to allow fitting of pantographs and a raised observation area to support overhead line testing.
Test Car 6	Used in the main for locomotive testing, or where the test involved incorporating a test coach within a heavy freight train; also freight vehicle testing from 1997.
Test Car 10	High Speed test coach, used for various coaching stock and locomotive tests from 1982.

Table 4 - Test Car Purpose

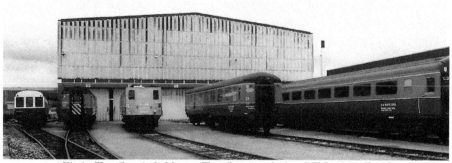

Fig.6 - Test Cars 1, 2, Mentor, Test Cars 6 and 10 at RTC, 1987 [Serco]

Other vehicles in the DM&EE fleet during the 1980s and 1990s included Test Car 4 and Test Car 5, both of which were originally converted from passenger vehicles for the British Rail Research Division (BRR), subsequently being transferred to the DM&EE when the BRR downsized its fleet of Laboratory Coaches.

TEST CAR 1 – ADW150375

Hawksworth designed auto trailer number W233 was one of fifteen examples which were built at Swindon in Lot 1736 to Diagram A.38 in 1951. After withdrawal from passenger service the coach was used between 1965 and 1968 by the regional Chief Mechanical & Electrical Engineers (CM&EE) department in Swindon as a base for engineers investigating running problems with freight wagons. During this period most of the seats were removed and, in their place, installed along the length of the main saloon area were battery packs which were used to power the simple test equipment; although the use of batteries limited the length of test runs because they had to be re-charged by plugging into a workshop charger. In 1968 when the regional CM&EE departments were centralised to form the DM&EE department, the coach was transferred to the Railway Technical Centre (RTC) in Derby and renumbered to ADW150375. At this time the coach still fitted with its original Great Western design leaf spring bogies, instrument batteries and painted in British Rail Maroon livery. Initially the DM&EE test division continued to use the coach in much the same way, although the charging system for the batteries was rigged up to run off the underframe mounted belt driven dynamo, which allowed the duration of test runs to be increased. A simple speedometer was fitted, deriving the speed from an optical sensor picking up the wheel rotation from a white painted line on the side of the wheel.

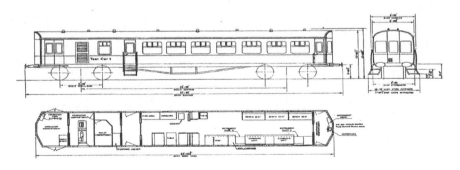

Fig.7 - ADW150375 Test Car 1 General Arrangement, 1978 [Serco]

Test Car 1	Vehicle Details
Original Vehicle	W233, LOT 1736 DIAGRAM A.38, Hawksworth design
Built	Swindon, 1951
Converted	RTC Derby, 1973
Weight	Generator end - 16 TONS 18 CWTS, Instrument end - 16 TONS 00 CWTS
Wheelsets	3' 6" Diameter, Journals 8" x 4" - Plain bearings, (from 1981 - Roller bearings)
Brake System	Dual braked & hand braked
Length	Over buffers - 67' 0", Over body ends - 63' 0³/₄", Bogie centres - 43' 6"
Width	Over step-boards - 9' 0", Over Steps (extended) - 10' 10"
Bogies	Original - GW Type 9' 0" wheelbase, (from 1981-BR standard B4 Type)

Table 5 - Test Car 1 vehicle details

The original auto trailer extending steps under the centre door were retained; being operated by levers inside the centre vestibule one for each side; the handrails were also connected to the sides of the middle step so providing an extended safe hand hold when climbing in or out on the occasions when the steps were folded out for use. Around 1970 the coach was painted in the DM&EE Blue & Red livery and designated Test Car 1 (TC1). The instrumentation systems were becoming more advanced requiring more power and instead of the batteries for equipment power, a small hand cranked generator was temporarily installed in the luggage area with the exhaust pipe fed out of an open side window when in use.

Fig.8 - ADW150375 Test Car 1 at the RTC, August 1987

Operation of early test trains including air braked wagons were initially a problem because TC1 was only vacuum braked, therefore to cater for such occasions a through air pipe was installed and test trains were marshalled using additional brake power vehicles in the form of two redundant Southern Region (SR) Mk 1 air braked coaches. Freight vehicle test trains at this time would have been operated to a maximum speed of around 60 mile/h.

Fig.9 - ADW150375 Test Car 1 at the RTC Way & Works, February 1987

Between 1973 and 1976 conversion work at Derby was undertaken to TC1, this work included installation of a two-pipe auto air brake system whilst also retaining the original vacuum braked system; air and vacuum brake pressure gauges were fitted into a specially designed mounting box inside the observation end area above the end windows. Two-tone electrically powered warning horns were installed on both vehicle ends operated by lever switches above the observation end windows and in the workshop area. A number of first class seats, tables, electric heating units, a toilet and an internal bulkhead which were recovered from a one of the SR Mk 1 first class coaches were installed; and an axle mounted tachometer for driving a more reliable independent test car speedometer. The recovered polished wooden bulkhead, which had windows and a sliding door, was fitted in the main saloon instrument area and provided a separated area for the kitchen the length of two window bays in front of the centre vestibule. The kitchen area was fitted out with two pairs of the Mk 1 coach first class seats, a table to the right side, and cupboard units constructed on the left side to include a sink. Installation of the toilet was to the rear of the centre vestibule and behind that a fire-proof generator room housing a new Perkins 3-cylinder, 240 volt diesel generator set with an exhaust outlet through the roof and bodyside radiator grille; diesel was supplied from an underframe mounted fuel tank.

Fig.10 - ADW150375 Test Car 1 at the RTC, May 86

A control panel for the new generator including electrical circuit breakers, the fire extinguisher system control and fire alarm bell were installed on the kitchen side of the vestibule bulkhead. The old luggage compartment at the very rear of the coach became a workshop area with a workbench, vice and tool storage. In place of the original Great Western design of oval buffers, large 22" diameter round head Oleo pneumatic buffers which were recovered from a first generation Diesel Multiple Unit (DMU) vehicle were fitted; this entailed modifying the bulkhead with extended buffer mounts. The reason for fitting the larger buffers was to minimise the risk of buffer locking or over-ride, whilst coupled to rolling stock with a large variety of buffer head sizes and extremes of buffer heights due to vehicle load condition. Instrument cable access was installed at the observation end of the coach just under the centre lamp bracket. The cables connecting to the instruments fitted to the test wagons were fed through the access hole direct to an inside plugboard installed just under the left side observation windows which was hard-wired through an underfloor trunking direct to the rear of the instrument racks that were installed along the right hand side of the main saloon area. In between test jobs effective use for the old GWR centre lamp bracket was often made for holding a loom of instrument cables awaiting fitting to the next freight vehicle under test.

The coach was designated TOPS code QXX in 1974, and continued to be used for freight vehicle testing based at the RTC Derby until 1997. Other modifications followed in 1981 during a period when TC1 was out of use for a general repair, including the fitting of more modern B4 bogies, which were also recovered from the SR Mk1 coaches that enabled the maximum operating speed to be increased to 100 mile/h.

Fig.11 - ADW150375 Test Car 1 instrument area, 1986 [Serco]

The Great Western Chocolate & Cream repaint, (including the gold-leaf lettering, lining and the Great Western Crest), was applied at the RTC EDU Workshop in 1985 as part of the Great Western 150th anniversary celebrations; although the coach was never actually painted in this livery during its passenger service days with British Rail. Like the other test coaches, the wording "Scientific Instruments Shunt with Care" was stencilled on the vehicle ends, and the mandatory "Danger Overhead Live Wires" warning labels applied. After the repaint at some point I also remember the carpet fitting in the instrumentation saloon area (probably sometime in 1986), the carpet was the deep red and black striped style used in Mk II coaches and made a huge difference to the warmth and general comfort inside the coach during testing. Other than that there were not many major changes to the coach after 1985, however various other small modifications were carried out over the years including fitting of a camera for viewing wagon wheelsets whilst running (located under the headstock), instrument area curtains in the windows, observation end screen wipers and two air conditioning units on the left side of the main instrumentation saloon area. Until the early 1990s, the instrumentation systems used within the test coach recorded test data that was subsequently passed back to the RTC office based analysis laboratory for post-test evaluation. The development of an on-board analysis system speeded up significantly the time it took to provide a formal test report; in some cases the result graph evidence of the vehicle under tests running characteristics could be printed on-board at the end of each test run. Between 1985 and 1999, I worked on dynamic test runs in Test Car 1 on over two hundred and twenty occasions.

A more detailed insight into the on-board testing activities is covered in later chapters of this book.

Test Car 1 was retired from test train duties in late 1997, and with the help of the 13809 7F Preservation Group in 1998 was moved to the Midland Railway Centre (MRC).

Fig.12 - ADW150375 inside the museum at the MRC in 2005

After storage in the MRC sidings covered with a tarpaulin for about a year, it was repainted into its original British Rail Maroon livery and for a period was housed within the Matthew Kirtley Museum building at the MRC Swanwick Junction site. The coach was subsequently purchased by the Locomotive 5542 Limited preservation group, and moved on to the South Devon Railway at Buckfastleigh in mid-2013, where the coach has regained its original identity W233 and is now restored for use again as an auto-trailer in preservation.

TEST CAR 2 – ADB975397

This test coach was first used by the DM&EE in 1974, after the original Test Car 2 was damaged during a test, and subsequently withdrawn from use and scrapped. The replacement coach was built in 1962 at Wolverton in Lot 30699 as a British Rail Mk1 BSK number 35386 and fitted with Commonwealth bogies. The coach was temporarily adapted for use by the DM&EE and renumbered ADB975397 in 1974, receiving various further modifications in 1976 to become the second generation Test Car 2 (TC2) with a TOPS code of QXX. Test Car 2 was predominantly used for slip/brake testing, being a regular visitor to the normal slip/brake test site on the West Coast Main Line between Crewe and Winsford.

Modifications included installation of a Perkins 3-cylinder 20 kVA diesel generator providing 240 volt AC power for on-board instrumentation systems, an underframe mounted 90 gallon fuel tank, heating, lighting and cooking facilities. A workshop area was provided along with a fitted kitchen, toilet, full height instrumentation racks were permanently installed in the instrument area, with table areas and storage cupboards.

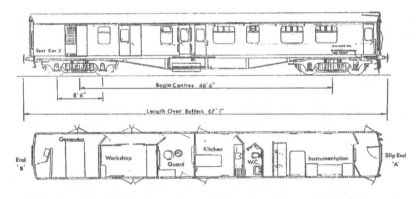

Fig.13 - Test Car 2 General Arrangement, 1988 [Serco]

The working test area end of the vehicle (referred to as the Slip-End), was fitted with a large gangway end inward opening door with a large drop-down viewing window that provided the working area platform for operating the slip/brake tests. Additional footsteps were also added under the slip-end right side door to assist climbing in and out during testing. Twin electric air horns were fitted at both ends of the coach operated by ceiling level interior levers. Additionally at the slip-end of the coach an air horn was fitted, fed with main reservoir pipe air via an operating lever valve to provide a louder warning that was used by the Officer in Charge (OIC) when undertaking slip/brake testing propelling moves.

Fig.14 - ADB975397 Test Car 2, A-end (Slip-end), July 1987

During the period between 1988 and 1990 several further modifications were made to TC2. The bodyside doors at the slip-end of the coach were made inward opening in October 1988, and twin full height grab handrails added using reclaimed parts from the guards compartment of a withdrawn Mk 1 coach. These modifications made getting in and out of the vehicle, particularly during slip/brake testing a lot easier and safer. The duplicate brake pipe and main reservoir piping arrangements on the headstock at the slip end of the coach were also modified to make things easier for coupling to the variety of freight wagons tested.

Fig.15 - Test Car 2 'B' generator end at the RTC Derby, July 1987

The installation of additional body end windows in September 1990 at the slip-end of the coach was primarily to assist the OIC with visibility of the vehicles during slip/brake testing. In 1991 further underframe piping modifications were made and additional isolation cocks were installed to enable the operation during slip/brake testing with air-braked locomotives rather than the original set-up that relied on the locomotive and test car brakes being operated by vacuum. During the mid-1990s the by then redundant headstock vacuum brake pipes were removed and sealed, although the underframe vacuum pipes and the vacuum brake cylinders remained in place. Another very useful modification to be fitted to TC2 in the early 1990s was the installation of an electrically driven 10 bar air compressor and a dedicated 150 litre air reservoir. These were installed in the rear of the vehicle behind the generator and rated to provide air at 7 bar for feeding the main reservoir pipe and through the reducing valve on the coach to control the brake pipe; thus enabling static brake tests to be carried out on coupled vehicles without the need for a separate air supply from a workshop or a locomotive. In late 1992 the coach was painted in the old British Rail lined maroon livery complete with roundel crest, although wasp stripes were added to both ends to assist with visibility.

Fig.16 - Test Car 2 Interior during Class 92 Tests, November 1994

The majority of the test projects on which TC2 was used were braking type tests, including slip/brake tests, for which the amount of instrumentation and recording equipment was not normally great. On the small number of occasions when the coach was used for tests involving larger amounts of instrumentation, then additional instrumentation racking was installed; an example of this was the Class 92 locomotive testing programme during 1994, where ride, brake, air pressures and pantograph performance were assessed during test runs on the WCML and through the channel tunnel.

Fig.17 - Test Car 2 at Winsford, February 1986

A fixed high intensity headlight and a tail light were installed at the slip end of the coach during 1997, primarily for use during slip/brake testing. The large round brake pressure gauges mounted above the bodyside window in the instrument area showed vacuum level, main reservoir pipe pressure, brake pipe pressure and brake cylinder pressure. Instrument cable access was provided at the slip-end of the coach on the right side; the instrument cables routed from the transducers fitted to vehicles under test were fed through a specially made access hole on the underside of the body just behind the side door.

Fig.18 - Test Car 2 in British Rail maroon livery, 1992

On the inside of the coach in the vestibule area there was a 48 connection plugboard where the incoming cables could be connected. The rear of the plugboard was permanently wired with cables fed into a short section of underfloor trunking into the instrument area up to the rear of the instrument racks where they could be connected as required to the inputs of the signal conditioning amplifiers. Between 1985 and 1999 I worked on slip/brake test days in TC2 on approximately 140 occasions, this gives some idea of the usage that this coach had, making the transit trip on each occasion between Derby and Crewe then slip/brake test working along the down slow line between Crewe and Winsford before returning to Derby. In February 2011 the coach was retired from Derby based test operations and transferred by road to the Old Dalby test facility by Serco, where it was used occasionally in movements up and down the test line until 2016 when it was transferred again by road to the Great Central Railway for preservation.

Fig.19 - Test Car 2, Serco Livery, head and tail lights installed

TEST CAR 3 - ADB975091

Mentor was converted at Swindon in 1973 from a standard British Rail Mk 1 BSK coach number 34616 of 1955 vintage and numbered ADB975091. The more modern B4 type coil sprung bogies were fitted to enable the ride performance to be improved from that of the original BR1 type leaf sprung bogies, which was necessary to support more accurate measurements of the overhead line. It was transferred from being a regional operated vehicle to the DM&EE fleet in the 1980s and was designated as Test Car 3 although it was not very often referred to as such.

Fig.20 - ADB975091 Test Car 3 at Motherwell with 37184, June 1985

The MENTOR coach was always hauled by a diesel locomotive whilst testing, this was because the use of an electric locomotive, with its pantograph raised, would have caused overhead wire disturbance ahead of the instrumented pantograph affecting the validity of the results. The main instrument area was at floor level in the saloon, however a raised section high level observation area provided the test engineers with an excellent view of the pantograph and overhead line from the roof area windows. A toilet, a small kitchen area and a diesel generator were installed to provide power for the coach heating, lighting and instrumentation systems, including the roof area lighting to allow the overhead line tests to continue in low light conditions. I did not get many outings on Mentor but took the opportunity when it arose to gain some experience of a very different side of the DM&EE testing activities during the 1980s. At the time of writing MENTOR was still in operation as part of the Network Rail infrastructure monitoring fleet.

Test Car 6 – ADB975290

Test Car 6 (TC6) was converted in the mid-1970s at the RTC Derby from one of the first production order British Rail Mk II FK vehicles (S13396), being renumbered ADB975290 with TOPS code QXA. The coach was predominantly used for locomotive testing being fitted with heavy duty drawgear and B4 bogies; being C1 gauge enabled it to negotiate many sidings, loading bays, and engineering works that the larger C3 gauge Test Car 10 could not. The interior was laid out with a large instrument saloon area incorporating instrument racking and bench working areas, a kitchen and twin diesel generators housed within fire-proof enclosures.

Fig.21 - ADB975290 Test Car 6 at the Railway Technical Centre, May 1986

There was also a toilet (retained from the original vehicle) and the test car had windows fitted within the vestibule end gangway doors; it was through wired with associated jumper cables mounted on the headstocks to allow coupling in High Speed Train (HST) train formations. A 27-way 'blue star' locomotive control cable was fitted and a temporary regulating air supply pipe was also installed on occasions to support coupling of locomotives to both ends of the coach. Twin electric air horns were fitted at both ends and instrument cable access at the instrumentation saloon end of the coach. The instrument cables entered slots just above each buffer in the coach end, then being fed to a plugboard which was connected via underfloor trunking direct to the rear of the instrument racks. The twin generators were replaced by a single larger generator unit in the mid-1980s and the associated external radiator grilles also modified. The instrumentation data recording capacity was significantly increased at the same time by the addition of a second instrument cable plugboard.

Fig.22 - Interior of Test Car 6 [Left] 1986, [Right] 1999

In 1991 the coach ends were painted yellow with wasp stripes applied and the blue/red livery retained until Serco took over in 1997 when their corporate livery was applied. By the late 1990s the interior instrumentation racks were replaced with more modern units and the vehicle end gangways had been removed as by this time it was used extensively for freight vehicle testing following the retirement of Test Car 1.

Fig.23 - ADB975290 Test Car 6 at the RTC in Serco livery, 1999

After a period stabled out of use at the RTC in Derby, the coach was withdrawn and by 2011 had been moved to Eastleigh pending asbestos removal, however in mid-2013 it was subsequently scrapped.

TEST CAR 10 – ADB975814

This Mk III vehicle was converted in the early 1980s at the RTC Derby from one of the prototype Inter-City 125 High Speed Train (HST) first class coaches, E11000/W41000. In 1982 the coach was renumbered to ADB975814 and a slightly non-standard DM&EE blue/red livery applied, with a red instead of a blue band at cantrail level. As with all other Mk III vehicles, the coach was C3 gauge, it was allocated with TOPS code QXA and designated Test Car 10 (TC10). Its conversion for use as a test coach included the installation of a large 3-Phase diesel generator unit and associated electrical control system to provide power for the instrumentation and also the coach air conditioning system.

Fig.24 - ADB975814 Test Car 10 at Darlington, February 1989

The extra length of the Mk III coach body compared to TC6 meant that the accommodation and test area installation could be much larger accommodate more comprehensive instrumentation installations.

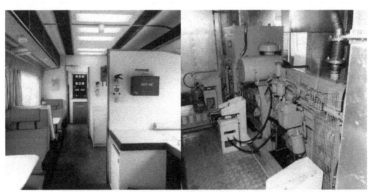

Fig.25 - Interior of Test Car 10 [Left], Generator [Right]

The layout of the main saloon instrumentation area was similar in layout to TC6, with instrument racks along the length of the saloon with intermediate tables / work areas and an aisle behind the instrument racking to allow easy access for connecting cables to the equipment. A separate kitchen area was provided with a sink, fridge, microwave oven, baby belling cooker and storage cupboards. A number of internal store areas were provided, including cloak cupboards and the original toilet was retained at one end. Externally the coach was fitted with drop-head buckeye couplings, buffers and drawgear, it was through wired with headstock jumper connections to enable it to be coupled within production HST formations.

Fig.26 - Test Car 10 behind 43051 with Mk IV test train at Heaton, February 1989

The coach was used for a variety of locomotive, coaching stock and occasionally freight vehicle dynamic tests from 1982 until 1998 when it was withdrawn from test train use, being saved for preservation and moved to the Peak Rail in Derbyshire it was stored for a period at their Matlock station site. By 2003 the coach was acquired by Network Rail, re-registered for use on the main line again, refurbished and modified for use in the High Speed Train (HST) based Network Rail New Measurement Train (NMT) where it is still in operation at the time of writing.

OTHER DM&EE TEST VEHICLES

Throughout the 1980s and 1990s the freight test team used a 2-axle van (ADC201055) that was often seen attached to the non-observation end of TC1 during dynamic test runs. This wagon was originally a standard British Rail Railfreight (TOPS code VDA) 24.5 tonne capacity, 20ft 9inch wheelbase van, design code VD001C built in Shildon in 1976 under Lot 3856. It was one of a batch of one hundred VDA vans fitted with disc brakes, a taperleaf spring design (designated FAT13 Taperlite type suspension) and a designed maximum operating speed of 75 mile/h. The wagon was transferred to the DM&EE test department in the early 1980s and designated TOPS code ZXA. Other than the fitting of test instrumentation brackets and cabling for connecting the locomotive intercom system during test runs, the van did not receive any modifications for use by the test section. The van also provided secure storage space for spare instrumentation racking and tables that were used in temporary test coach installations.

The inclusion of the ZXA within the often short freight vehicle test trains provided additional brake force, but the main purpose for its use was as a ride comparator vehicle, and a benchmark for the assessment of ride performance of new or modified freight wagons and on-track plant. The consistent riding properties of the Taperlite suspension fitted ZXA also helped in establishing locations of discrete or unrepresentative track features that could affect the assessment of the vehicle under test. The ZXA vehicle was saved for preservation following withdrawal from test train use in 2002, and at the time of writing is residing at the Great Central Railway Nottingham.

Fig.27 - ADC201055 at the RTC in original brown livery, February 1987

The first generation DM&EE Test Car 4 was the 1960s British Rail Western Region Dynamometer Coach DW150192, which was no longer in use by the DM&EE when I joined the test section in 1983. The second vehicle that became designated Test Car 4, numbered RDB975984, was a Mk3 Kitchen Buffet coach built in 1973 in Derby to Diagram 42, in Lot 30849 as part of the prototype HST numbered E10000; it was converted in the early 1980s to Lab Coach 15 'Argus' for the British Rail Research Division. In the early 1990s it was transferred to the DM&EE fleet and designated Test Car 4 being used on a variety of vehicle acceptance test programmes including some freight vehicle tests. The coach survived following withdrawal from test train use in the early 2000s, and along with TC10, was acquired by Network Rail, refurbished, and modified for use in the NMT where it is still in operation at the time of writing.

Fig.28 - Test Car 4 RDB975984 in Railtest livery, 1995 [Serco]

The original DM&EE Test Car 5 vehicle numbered ADB975051 was converted from a British Rail Mk 1 BSK coach and was used in conjunction with the Advanced Passenger Train (APT-P) power car testing project. This vehicle was retired in the late 1980s and used for a period in the Crewe Works test train with its windows plated over before moving to Peak Rail and preservation. The second generation Test Car 5, RDB975422 was originally converted at Swindon in 1978 from a Mk 1 BSK coach W34875 for British Rail Research Division as Laboratory 6 (Prometheus). The vehicle which had a flat roof area at the instrumentation end for pantograph installation, was fitted with BT10 bogies and had a maximum operating speed of 125 mile/h. It joined the DM&EE test car fleet in the late 1980s, was re-designated as Test Car 5, and was used occasionally for test duties; however more often it was seen as an additional vehicle in test trains providing brake force. The coach was stabled out of use for a few years at the RTC sidings before being scrapped, it was finally moved away by road in 2007.

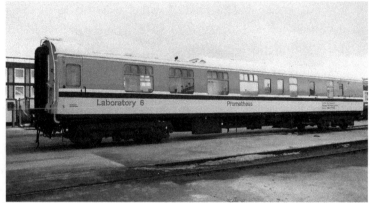

Fig.29 - Pre-Test Car 5 days, as Lab 6 RDB975422 at the RTC Derby, July-1987

TEST TRAIN LOCOMOTIVES

The locomotives that were used during the 1980s and 1990s for hauling DM&EE freight and on-track plant vehicle test trains varied considerably. My early experiences working on the vehicle testing section included encounters with Class 20, 31, 37 and Class 45 locomotives, however from the mid-1980s onwards for the majority of diesel hauled test trains, Class 47 locomotives were used. The regular test train operations for the majority of freight and on-track plant vehicles were undertaken on the Midland Main Line between Derby and Bedford or Cricklewood sidings; and also between Derby and Crewe providing access to undertake the slip/brake testing on the level gradient of the down slow line between Crewe and Winsford.

Fig.30 - Class 45, 45041 at Crewe with Test Car 2, July 1987

On occasions when undertaking slip/brake testing of heavier vehicles, or when test speeds above 75 mile/h were required, an electric locomotive was attached at Crewe (Class 85 or Class 86) to provide the necessary higher acceleration performance required to get the test trains up to speed within the relatively short slip/brake test site distance.

Fig.31 - Class 20s, 20208 + 20173 at Wigston Junction, June 1989 [Mike Fraser]

Examples of more unusual workings were on the 28th November 1987 and the 20th June 1989 when Class 20 locomotives were used to haul test trains. The first example when 20097 was used to haul the slip/brake test transit movement from Derby RTC to Crewe. The Class 20 unusually operating singly, nose end first; on this occasion the Class 20 was replaced with a Class 47, 47113 for the purposes of testing between Crewe and Winsford. The ride test in June 1989 saw a pair of Class 20 locomotives operating in the more usual formation with nose ends coupled together hauling Test Car 1, the ZXA van and two laden condition 102 tonne O&K hopper wagons running from Derby to Bedford via Corby and returning via Leicester.

Fig.32 - Class 20, 20097 at Crewe, November 1987

For the purpose of freight vehicle testing certain locomotive types were more suited, and others less so. In general the Class 45 and Class 47 locomotives were preferred for the ride test runs on the Midland Main Line route where the running speed needed to be controlled quite accurately within speed bands during the tests. These locomotives were preferred because the responsiveness of the engine control system of the Sulzer powered locomotives (compared to the English Electric type locomotives) made it easier for the driver to control the speed more accurately with the short and often light trailing load of a test train.

LOCOMOTIVE ZXA TEST CAR 1 VEHICLE UNDER TEST

Fig.33 - Typical freight vehicle ride test train formation

The normal practice was for the outward test run to leave Derby with the vehicle under test trailing, loose coupled to the observation end of the Test Car. The locomotive was run-round to the other end of the train at the outward destination, and loose coupled to the vehicle under test for the return trip back the Derby. The reason for loose coupling was to ensure representative running conditions and the vehicle under test was not artificially stabilised by tight coupling to the adjacent test coach or locomotive. In the early 1980s when Derby '4 Shed' had a large allocation of locomotives, the traction for most test runs was easily provided from the plentiful supply of locomotives on hand at Derby; however as the High Speed Trains (HSTs) were introduced onto the Midland Main Line and locomotives started to be withdrawn, the provision of test train locomotives came predominantly from Toton and Crewe Diesel depots.

Fig.34 - Class 47, 47131 at Merehead, August 1986

Another rare occurrence of an English Electric locomotive hauling a ride test train was seen on the 10th August 1987 when Class 37 number 37222 was provided from Toton for the ride test run of a new container flat wagon between Derby and Finedon Road (Wellingborough). Class 31 locomotives also appeared now and again instead of the more usual Class 47s to haul ride test trains on the Midland Main Line.

Fig.35 - Class 37, 37222 at Glendon North Junction, August 1987

In August 1987, Class 31 locomotive number 31107 was provided to haul a ride test train run between Derby and Cricklewood, little did we know then that this locomotive would become a star on screen after its withdrawal. In 2006 it was used in a campaign to raise safety awareness at level crossings, and was deliberately crashed into a car at 70 mile/h, the film also being aired on the Top Gear motoring television programme. During 1986 a Class 31 locomotive number 31298 was allocated to RTC test train duties, unfortunately the locomotive, which was renumbered 97203 and re-painted in RTC livery, did not last long in this role, being withdrawn in 1987 due to fire damage. It was replaced by another Class 31 locomotive number 31326 in 1987, which was re-painted in a similar RTC livery and renumbered 97204; this locomotive was subsequently renumbered again as 31970 in July 1989.

Fig.36 - Class 31, 97204 at Winsford, June 1989 [Mike Fraser]

This Class 31 was often seen heading Research Division trains; however it was rarely used for the DM&EE test trains, an exception being at least one occasion for hauling a slip/brake test with O&K hopper wagons between Crewe and Winsford in late June 1989, although I was not aboard the test train that day. Apparently the amount of slip/brake tests achieved during the running on the 22nd June 1989 was less than normal due to the Class 31 being somewhat lacking in power, which was more noticeable because the two 102 tonne O&K hoper wagons being tested were fully laden.

Fig.37 - Class 31, 31415 at Corby, April 1988

In 1988 four Class 47 locomotives based at Crewe Diesel depot were allocated specifically to RTC test train duties and were subsequently renumbered in the 97xxx series to denote this. The application of the 97xxx number series only lasted for a year or so, and three of the RTC allocated locomotives were renumbered again in 1989 with 479xx series numbers applied, being joined by three more during 1990 and 1991.

The six Class 479xx locomotives were painted in various different liveries. including large the standard BR blue, both with and without a large double arrow logo, departmental grey, civil engineers grey & yellow, InterCity Mainline, British Rail Technical Services livery and a Midland Counties special maroon livery; all six also received nameplates as noted in the table below.

Previous TOPS Numbers	Renumbered 1987-1988	Renumbered 1989-1991	Livery	Nameplates
31298	97203	-	BR Research	-
31326	97204	31970	BR Research	-
47480	97480	47971	Blue (Logo)	Robin Hood
47545	97545	47972	Blue (Logo) & T.Services	The Royal Army Ordnance Corps
47034, 47561	97561	47973	Blue & Maroon & InterCity	Midland Counties 150 1839-1989 & Derby Evening Telegraph
47472	97472	-	Blue	-
47531	-	47974	Grey/Yellow	The Permanent Way Institution
47540	-	47975	Grey/Yellow	The Institution of Civil Engineers
47546	-	47976	Grey & Grey/Yellow	Aviemore Centre

Table 6 - Class 31 & 47 locomotives specifically allocated for use on test trains

The fourth 97xxx locomotive (97472) reverted to normal duties during May 1989, carrying its original number 47472, and was subsequently withdrawn in February 1991 after colliding with another Class 47 locomotive (47533) at Reading.

Fig.38 - Class 47, 47545 at Crewe, June 1988

All of the Class 47 locomotives allocated to RTC duties featured in hauling the DM&EE testing section ride test and slip/brake test trains from 1988, and were also used for hauling many other test trains and infrastructure monitoring trains such as the Mobile Electrical Network Testing, Observation and Recording coach (MENTOR), and the Structure Gauging Train (SGT).

The RTC resident ex-Class 24 locomotive number 97201 was, as far as I know never used for the purposes of hauling DM&EE ride or slip/brake test trains, being allocated to the Research Division and primarily used to haul the Tribometer Test Train. I did however, pass the locomotive on many occasions when accessing test trains during the 1980s seeing it, languishing in the RTC sidings awaiting its next outing, therefore I think it deserves a mention here.

Fig.39 - Class 24, 97201 at the RTC Way & Works sidings, July 1987

On occasions other types of traction were used to haul freight test trains where tests on the locomotives themselves, or representative train operations were required. For example some aggregate train tests on the Western region utilised Class 56 or Class 59 locomotives. The passenger stock testing for the development of the SIG bogies and Mk IV coaches used a variety of traction during the long and varied test programme which included operations on the West Coast Main Line (WCML) and East Coast Main Line (ECML). Locomotive types used for transit movements and testing included Class 31, 43(HST), 47, 81, 86, 87, 90 and Class 91.

Where I did keep the records of the locomotive numbers used on particular test trains, these have been logged for reference in the test log tables in the book appendices.

Fig.40 - Class 47, 47971 at Winsford, 1989 [Mike Fraser]

RUNNING CHARACTERISTICS

The assessment of the safe dynamic running characteristics of rail vehicles considers the interface between wheel and rail during dynamic running that can be established by investigating the forces present at the wheel to rail interface. The following four key areas may be considered when undertaking acceptance testing of new or modified traction and rolling stock with respect to running characteristics.

- Safety against derailment, i.e. determination of an acceptable level of the relationship between vertical and lateral wheel forces; the vertical wheel load 'Q' and the lateral force between wheel and rail 'Y'. This relationship is referred to as Y/Q (or Y over Q).

- Safety against track shifting, ΣY forces (The sum of the lateral forces at both wheels on an axle).

- Vertical wheel forces 'Q' must be maintained below permissible rail stress levels.

- Instability, normally assessed up to the maximum designed operating speed plus 10%.

Fig.41 - RTC Derby Weighbridge, Procor 40 tonne Conflat wagon, July 1986

The first step when assessing a new or modified vehicle design was to undertake static tests. When new rolling stock arrived at the RTC for acceptance testing, after initial inspection and recording of all relevant information, running numbers, suspension type and condition, relevant equipment serial numbers etc., the vehicle was moved into the RTC weighbridge; this was a specifically designed rail vehicle weighing facility with eighteen individual sections of rail, each incorporating calibrated load cells.

Each rail weighing cell had a capacity of 20 tonnes, and the spacing of the load cells enabled the simultaneous recording the individual loads (Q) of each wheel of most types of freight wagons, passenger coaches and locomotives. In cases where the wheel spacing was such that the wheels could not align with individual weigh rails, for example an 8-axle nuclear flask wagon with four bogies, then the vehicle had to be moved slightly between the weighing of each bogie. It was normal practice to weigh vehicles two or three times, with a shunt movement up/down the RTC yard in between each weighing. This was undertaken to evaluate whether any uneven wheel loads were due to the effect of hysteresis within the suspension or a true suspension set-up problem. The next step for static testing was to determine the potential for slow speed derailment, i.e. determination of an acceptable level of Y/Q; 'Q' being the vertical wheel load and 'Y' the lateral force between wheel and rail.

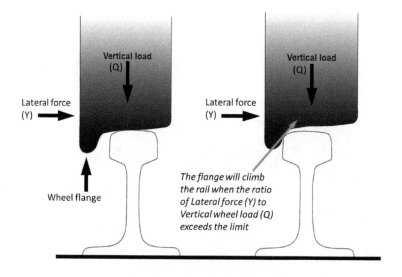

Work carried out by the British Rail Research Division established that to maintain safe running then the quotient Y/Q should not exceed 0.8. Computer simulations or a true measure of Y/Q could be validated or established by a number of methods, including by the use of instrumented wheelsets or instrumented rails in order to directly measure the wheel to rail forces; however these methods were expensive and often time consuming, therefore British Rail DM&EE developed a simplified (although considered conservative) method of demonstrating that Y/Q was within acceptable levels by using two much cheaper static tests and a simplified dynamic test process that could be undertaken relatively quickly by:-

- Measuring the torsional stiffness of the vehicle suspension / body.

- Measuring the bogie rotational resistance (for bogie vehicle types of vehicle).

- Measuring acceleration levels during dynamic running.

By the 1980s when I joined the DM&EE testing section, these methods commonly referred to as resistance to derailment, were well established and were used in the process of demonstrating safe running characteristics of all new and modified traction and rolling stock. The following sections are collated from my experiences and diary logs to provide in insight into the different stages of resistance to derailment testing and to show examples of how this was applied to a variety of types of rolling stock.

TORSIONAL STIFFNESS TESTS

The torsional stiffness test was often referred to as a Delta (Δ) Q/Q test, i.e. the change in wheel load (ΔQ) divided by the mean wheel load (Q) on any given axle. This test was undertaken by applying a twist to the vehicle body and suspension to simulate a defined level of dip in the track on one rail, whilst simultaneously recording the individual vertical wheel loads (Q). When a vehicle was twisted to simulate this defined level of track dip (1-in-300 acting over length of the vehicle, plus an additional 1-in-300 short wavelength twist acting over a distance of 6 metres), the resulting change in any of the wheel loads ($\Delta Q/Q$) had not to exceed 0.6, or 60% wheel off-loading. So basically, the suspension of a rail vehicle had to be flexible enough to allow the vertical wheel load to be maintained whilst traversing a dip in the track. A vehicle with a very stiff suspension or a rigid body/underframe was more likely to have a greater reduction in wheel load when traversing a track dip, therefore increasing the potential for derailment. The weighbridge at the RTC Derby had a feature built into one of the pair of weighing cells whereby the rail on each side could be hydraulically lifted or lowered. Therefore for establishing the $\Delta Q/Q$ for a two-axle vehicle or the short wavelength $\Delta Q/Q$ for a bogie of a vehicle, the hydraulic lift/lower equipment was used to simulate a track dip of the required level.

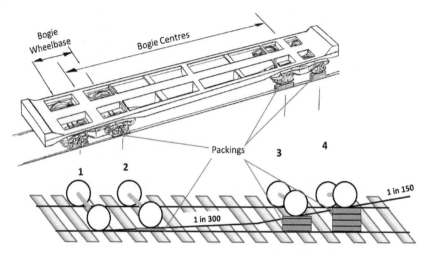

Fig.43 - Torsional stiffness test, bogie vehicle, jack & pack method

When it came to generating the required 1-in-300 track twist for a bogie vehicle such as a container flat wagon, then a manual method of jacking the vehicle and applying shim/packings between the wheel and the weighbridge load measuring rail was used. An example is shown below in Fig.44 of a vehicle undergoing a torsional stiffness ($\Delta Q/Q$) test at the EDU weighbridge in the RTC Derby on the 8th August 1987. The twist due to the vehicle underframe flexibility is noticeable in this view. During a torsional stiffness ($\Delta Q/Q$) test the brakes were always isolated to ensure they remained released to prevent the brakes affecting the wheel load readings; therefore wooden wheel chocks were installed either side of the non-lift wheelset to prevent vehicle movement. For additional safety, purpose designed bars were fitted between the wheel flange backs in line with each axle of a wheel to be lifted, locating along a special bar in the EDU weighbridge between the rails, this mitigated the risk of the wheelset sliding sideward during the test.

Fig.44 - Tiphook 82 tonne wagon TIPH93243 undergoing a torsional stiffness test

A torsional stiffness ($\Delta Q/Q$) test was also used to establish the level of hysteresis or friction content within a vehicle suspension; which was important when assessing the amount of damping available from a friction damped type suspension before embarking on a dynamic test. In extreme cases on vehicles with friction type dampers where the level of damping was much less than the predicted values, it was necessary to investigate further by inspection.

Fig.45 - Tiphook 82 tonne wagon GPS bogie during a torsional stiffness test

Often the cause of low levels of friction damping was attributed to oil or grease contamination on the friction surfaces, which meant that the bogie or suspension had to be dismantled to clean and degrease the friction damper surfaces. The facilities in the EDU workshop enabled this sort of work to be undertaken easily without the need to take the vehicle off-site for rectification.

Fig.46 - An example of a Y25 type 1.8m wheelbase cast frame bogie

An example of a load sensitive friction damping is shown in Fig.47. As the primary spring load increases, the plunger force increases proportionally via the sliding friction wedge, therefore the force exerted by the plunger squeezes the axlebox providing damping to a level proportional to the vertical load in the suspension. The primary suspension springs fitted to vehicles that carry large payloads often had two springs with different rates to provide for acceptable running characteristics when in the different load conditions. When the vehicle was in the tare (or empty) condition the outer primary spring was in use, and when loaded and the suspension compresses the slightly shorter inner higher rated spring comes into contact. It was normal practice to undertake torsional stiffness ($\Delta Q/Q$) tests in the tare condition only, this was because the worst case wheel off-loading was prevalent with lower axleloads; in the case of vehicles fitted with two stage suspension springs that could be partially loaded, a check of the torsional stiffness properties was also carried out in the load condition that aligned with the contact of the inner primary springs.

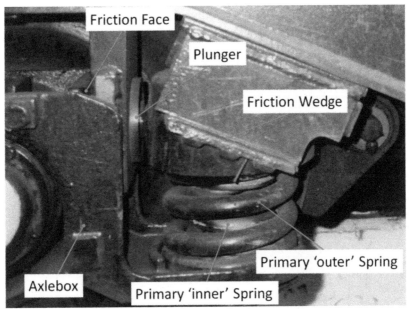

Fig.47 - Friction damper and spring arrangement

An example of a GPS type bogie is shown below in Fig.48, in this case a prototype In-Line Tipper Wagon for Redland, number REDA28100, which was presented for static tests including bogie rotation and torsional stiffness ($\Delta Q/Q$) tests at the RTC Derby in September 1987.

Fig.48 - GPS type bogie

The results from a torsional stiffness ($\Delta Q/Q$) test were plotted onto a graph with axes of track twist and wheel load to provide the test engineers not only with a clear decision on whether the vehicle had met the 60% wheel off-loading criteria, but also additional information about in-built twist in the bogie or the vehicle body, and the level of hysteresis within the suspension.

Different designs of bogie could exhibit very different characteristics during torsional stiffness ($\Delta Q/Q$) testing, for example a pedestal suspension type bogie with coil springs and a load sensitive friction damping arrangement could have in the order of 3 to 4 tonnes of hysteresis present when the vehicle was in the tare condition; this can be seen in the example torsional stiffness ($\Delta Q/Q$) test results graph in Fig.49 indicated by the average band depth of the butterfly shaped result plot.

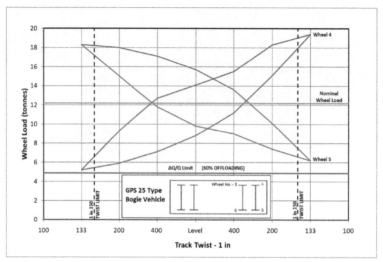

Fig.49 - An example GPS type bogie torsional stiffness test result

Newer designs of bogies incorporating hydraulic damping exhibited different characteristics when subjected to a torsional stiffness ($\Delta Q/Q$) test. An example of such a vehicle was the Low Track Force (LTF) bogie fitted ex BAA steel carrier wagon RDC921000 (designated YXA) in May 1987.

Fig.50 - LTF type bogie

The characteristics of the hydraulic damping and two stage suspension when tested statically presented much less hysteresis in the results which can be seen from the shape of the test results diagram plotted. In the example of torsional stiffness (ΔQ/Q) test results shown in Fig.51, the two metre wheelbase bogie had slight in-balance in the wheel weights when on level track. The amount of twist in the bogie or suspension set-up was established from the point at which the torsional plots for each wheel cross; in this case when approximately 1-in-400 twist is applied to the bogie.

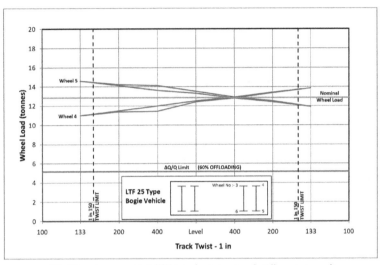

Fig.51 - An example LTF type bogie torsional stiffness test result

The 1-in-150 twist limit is equivalent to about 13.3 mm track twist on a 2 metre bogie, therefore proportionally the amount of twist in the bogie or suspension set-up can be evaluated to be in the order of 5 mm.

BOGIE ROTATION TESTS

The next stage of static testing following successful completion of the torsional stiffness (ΔQ/Q) test was to establish the bogie rotational characteristics on bogie type vehicles. These tests were conducted using the bogie rotation rig that recorded the force required to rotate a bogie of a vehicle whilst the bogie was rotated under the vehicle. The maximum angle of rotation of the bogie was calculated from the equivalent designed minimum radius curvature for the vehicle. The unique, specially designed test rig at the RTC in Derby, was installed in road number 3 within the EDU workshop. The rig had a rotating table on a low friction bearing to minimise the effects of the rig itself on the force measurements during testing. The rotating table was supported clear of the bearing when not in use and when shunting vehicles into place onto the test rig, this was to prevent damage to the bearing. This hydraulically operated test rig provided the capability to test vehicles with up to 25 tonnes axleload at bogie rotation angles up to seven degrees in each direction, at rotational speeds of up to one degree per second. The force measurements provided evidence for evaluation of the body / bogie yaw torque which is generated by the passage of the vehicle through curves, points and crossings etc. For main passenger lines the minimum curve radius was 120 metres, but for non-passenger lines and sidings this could be reduced to 90 metres, and in some cases privately owned sidings could have curve radii as tight as 70 metres.

Bogie rotation tests were normally conducted at two speeds of bogie rig rotation, 0.2 degree per second, and 1.0 degree per second. These angular velocities being representative of curve entry and exit in normal service conditions. For example the lower rate (0.2 degree per second) was representative of the rotation of a bogie when on the main lines traversing medium to large radius curves, whereas the higher bogie rotation rate was more representative of a vehicle traversing points and crossings. When evaluating the measurements, consideration was also given to the effects track geometry, and based on a nominal track gauge of 1435 mm (4' 8¹/₂"), gauge widening was applied as the radii of the curve decreased as follows:-

- on curves below 200 metres radius - 7 mm of gauge widening applied

- on curves below 140 metres radius - 13 mm of gauge widening applied

- on curves below 110 metres radius - 19 mm of gauge widening applied

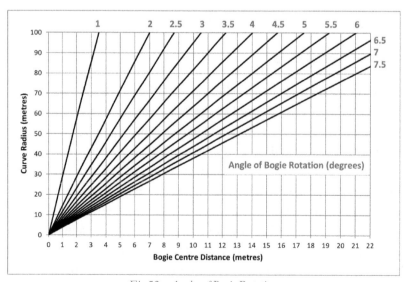

Fig.52 - Angles of Bogie Rotation

In addition to the measurements taken, an assessment of the clearances around the bogie and associated equipment was noted throughout the test. This was particularly important where electrical cables and air pipes were fitted between the body and the bogie of the vehicle, to ensure there was sufficient slack and free movement to cater for all angles of bogie rotation. The results of the bogie rotation tests were evaluated by calculating the unit-less bogie 'X' factor, which was the relationship between bogie yaw torque, wheelbase and axleload.

$$'X'factor = \frac{torque(kNm)}{axleload(kN) \ x \ wheelbase(m)}$$

The 'X' factor acceptance limit was 0.1 for a passenger vehicle, parcels vehicle or a locomotive, however for freight and on-track plant vehicles there was a dispensation below 8 tonnes axleload whereby the 'X' factor acceptance limit relaxed to 0.16 at 5 tonnes axleload.

Fig.53 - Container flat vehicle undergoing a bogie rotation test

Should the bogie rotation 'X' factor values fall between the design limit and the acceptance limit, it was normal practice to allow the vehicle into service but to request a re-test after a period of running. The recall of a vehicle for a bogie rotation retest was normally after completion of around 10,000 miles in service.

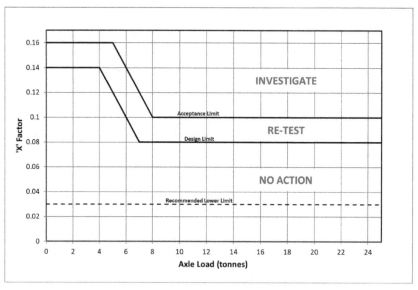

Fig.54 - Freight vehicle bogie rotation test limit 'X' factor values

Basically the higher the 'X' factor, the higher the lateral force (Y) that is generated during curving, and the increased potential for flange climbing and derailment. The lower the 'X' factor the less yaw control that the bogie has, which can lead to instability (sometimes referred to as hunting). Therefore optimum results were in the region of 0.6 to 0.8, depending on the vehicle and suspension type. Bogie rotational control was applied in a number of ways, for example friction sidebearers or bogie to body yaw control hydraulic dampers. Freight vehicles often had a flat or spherical centre pivot where the body locates onto the bogie with a friction lining material that provided some yaw control; however many vehicles relied on additional yaw control in the form of friction type sidebearers, often sprung to maintain a consistent contact force.

Fig.55 - Freight wagon UIC centre pivot [Left] & sidebearer arrangement [Right]

For passenger rolling stock operating at higher speeds, a greater degree of rotational control from hydraulic yaw dampers was required to provide additional bogie yaw control to prevent the onset of hunting which was more prevalent at higher operating speeds. Hunting predominantly occurs on straight track (i.e. with only small angles of bogie rotation), where unstable yaw oscillations of a bogie wheelset, or wheelsets become uncontrolled due to reactions between the wheel and rail forces.

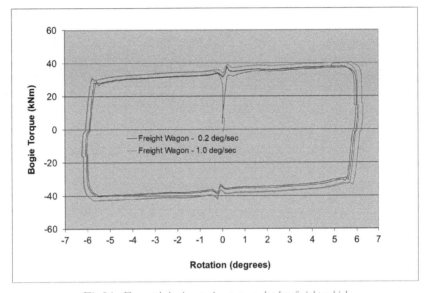

Fig.56 - Example bogie rotation test result plot, freight vehicle

On some vehicles the application of hydraulic yaw dampers to control hunting resulted in very high levels of yaw torque on smaller radius curves. This was alleviated by fitting of hydraulic yaw dampers that were designed to only be effective over part of their operational range. This provided greater yaw control on straight track and large radii curves, but less yaw control, therefore lower potential for derailment on tighter curves. During the 1980s measurements of the body/bogie yaw torque and bogie rig rotation angle were normally printed using an X-Y plotter during the test; this allowed the operator to evaluate the 'X' factor confirming acceptability immediately after each test was completed. As technology progressed in the 1990s, the measurements were also recorded onto cassette data tapes allowing post-test presentation into the test report. An example of a freight vehicle bogie rotation characteristic in Fig.56 shows the consistent yaw torque throughout all angles of bogie rotation. The rotational torque was also consistent for both slow and fast speeds of bogie rotation.

Fig.57 - Passenger coach undergoing a bogie rotation test

The example in Fig.58 from a passenger vehicle fitted with air suspension and yaw dampers shows how the rotational resistance increases with the angle of rotation, this was due to increasing stiffness as the air suspension twists.

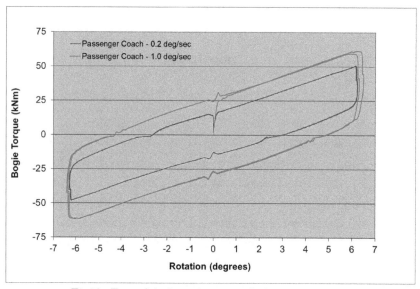

Fig.58 - Example bogie rotation test result plot, passenger vehicle

The effects of the different rotational speeds are also clear; the faster rotational speed giving a higher rotational torque, this was due to the yaw dampers fitted to control stability at the higher speeds. An unusual example of the use of a sidebearer arrangement for a purpose other than to control bogie yaw torque was seen fitted to the FAA container well wagons. The two different heights at each end of the sidebearer and the slope between can be seen in the photo in Fig.59 below.

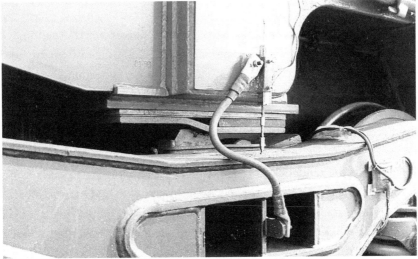

Fig.59 - FAA type wagon tilting sidebearer

The reason for this arrangement was to provide a small amount of tilt into the wagon underframe when traversing curves in order to maintain gauge clearance when the vehicle was loaded with 9' 6" high containers. The sidebearer arrangement did produce the desired tilting effect, however the wagons exhibited some less than ideal behaviour when traversing poor track geometry. Following a greater than normal period under test and additional investigation by the British Rail Research division, the wagons were eventually accepted for operation in service; however following a number of low speed derailments on poor track the titling sidebearer design was replaced with more conventional flat UIC type sprung sidebearers.

FREIGHT WAGON RIDE TESTING

Following successful completion of the static wheel weighing, torsional stiffness (ΔQ/Q) and bogie rotation tests, the next stage was normally to undertake on-track running tests as the method of assessing the dynamic running characteristics of a new or modified vehicle. The tests were undertaken by either a direct measurement of the wheel to rail forces using load measuring wheelsets, or by an indirect method using outputs from accelerometers mounted on the axlebox, bogie frame and underframe to derive the wheel to rail forces. Until 1977, simplified methods of assessing safe running characteristics of railway vehicles (particularly freight stock) were using measurements of accelerations in the vertical and lateral directions on vehicle headstocks. This method often referred to as 'Ride Testing' was usually based on manual assessment of chart traces by smoothing (filtering) of representative short 10 seconds sections of the acceleration time history records. The values of amplitude and frequency of the smoothed accelerations were then evaluated using pre-determined criteria to give a single figure result that was known as Ride Index. A maximum limit of Ride Index value of 5.0 at a frequency of 1 Hz and a mean limit of 4.25 at 1 Hz were set; however this methodology was derived originally for the assessment of passenger ride comfort, and therefore was not considered optimum for providing an indication of the tendency to derail. In 1977 a working party was set up within the DM&EE Test and Performance department with a remit to devise more suitable testing methods and analysis techniques together with appropriate standards (limits). The remit was set out guidelines to ensure that acceptance testing evaluated vehicle operational safely at its design speed and load condition over all normal British Rail track conditions; and to be safe against derailment under slow speed operation in private sidings. The Freight Group of the Test and Performance department went on to develop a new method for the assessment of static and dynamic running characteristics. This method developed used acceleration measurements retained from the Ride Index process, however the technique by which these were analysed was now automated, using a Peak Count Zero Crossing (PCZC) cycle counting process.

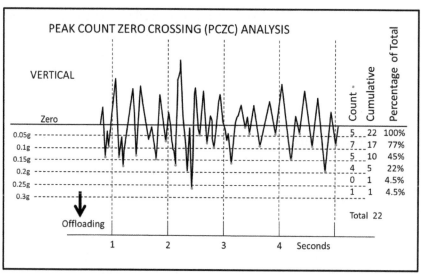

Fig.60 - Example of PCZC cycle counting method, vertical acceleration

The assessment and presentation of the results was classified based on constant speed test sections and the results for each speed band plotted on a percentage cumulative acceleration graph. The principle of the Peak Count Zero Crossing (PCZC) cycle counting method was an automated equivalent to the original Ride Index assessment of the amplitude and frequency of the smoothed accelerations. The acceleration signals were filtered at 6 Hz Low Pass (i.e. frequencies greater than 6 Hz were attenuated), the peaks were identified and classified into levels, normally $^+/_-$ 20 levels from the zero reference. For vertical accelerations the level interval was 0.05 g and for later accelerations the interval was 0.025 g. The number of peak counts in each level for each constant speed test section were counted and stored, then cumulated and presented as a percentage of the total counts for the respective test section. The concept of applying limits by an acceptance curve was established by the DM&EE working party, and took the form of a curve showing the permissible cumulative number of cycle counts above a given level plotted against that level as a percentage of the total counts for each constant speed test section. The limiting acceleration level curves were derived on an empirical basis from assessment of ride index records of a large number of freight wagon types that had varying performance records.

These limits became known as the Freight Acceptance Curves by which the assessment of the running characteristics was then undertaken for all new or modified traction and rolling stock until the late 1990s when the Railtrack Standards were developed. Ironically the Railtrack Standard applicable to the assessment of running characteristics for traction and rolling stock included a method of assessment based on the DM&EE developed Freight Acceptance Curve. This process was also carried over as a method of assessment into the Rail Safety & Standards Board (RSSB) standards from 2003 onwards. The Freight Acceptance Curves for the vertical and lateral accelerations were different. The vertical limiting curve was used for the assessment of only the off-loading cycle counts and incorporated a cut-off at 0.1 % of cumulative cycles counted for 0.65 g and above to recognise that occasional accelerations at such levels were attributable to inevitable track defects for which the vehicle suspension designer cannot be expected to cater. The lateral limiting curve followed the same principle; a 0.1 g cut-off was applied to cater for a situation where acceptable very mild oscillations could occur. A second cut-off at 0.1 % of cumulative cycles for 0.35 g and above was incorporated, again to recognise that occasional accelerations at such levels were attributable to track defects. These data recorded and presented on the acceptance curves was normally categorised in three separate ways for each test run:-

- the whole route for the vehicle maximum designed operating speed,

- the welded track sections for each constant speed test section,

- the jointed track sections for each constant speed test section.

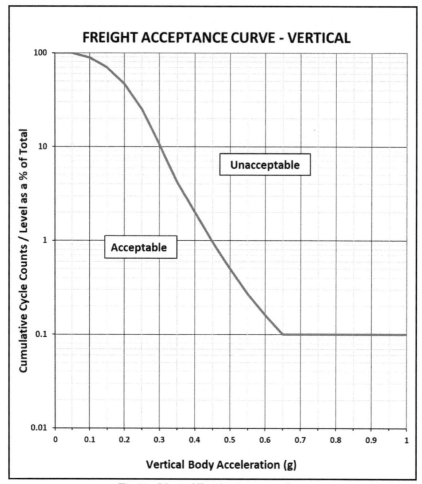

Fig.61 - Vertical Freight Acceptance Curve

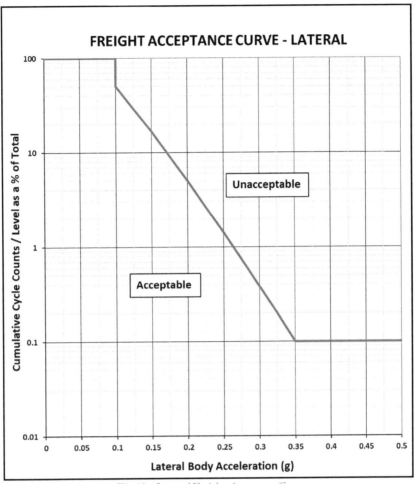

Fig.62 - Lateral Freight Acceptance Curve

The proportions of welded and jointed track sections over which data could be collected on a dynamic test varied between test runs due to operating constraints and also due to track maintenance and renewals; however a minimum distance for each test section of 10 miles was recommended to provide a satisfactory cycle count distribution for assessment using the PCZC percentage method. In addition to the cycle counting an assessment of the predominant or average frequency of each acceleration signal was undertaken. The time history below in Fig.63 shows the manual process for assessment of the predominant frequency. With the advent of the Freight Acceptance process this assessment of frequency was obtained for each constant speed test section by dividing the number of peaks on one side of zero for the test section, by the elapsed time in the test section. In the event that a suspension resonance or hunting was seen during a test run, then post-test evaluation including full frequency spectrum analysis was undertaken of the acceleration data to establish the exact frequency of the resonance at the specific point on the test run.

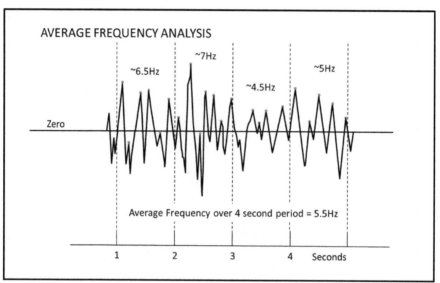

Fig.63 - Example of average frequency assessment method

A typical test instrumentation installation that was fitted to establish the dynamic behaviour of a bogie freight wagon included up to 20 transducers of various types that were selected dependent on the suspension arrangement of the vehicle under test.

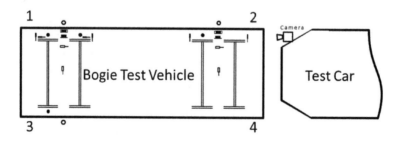

- ▬ Vertical Accelerometer (Underframe)
- ▲ Lateral Accelerometer (Underframe)
- ▮ Longitudinal Accelerometer (Underframe)
- ○ Vertical Potentiometer (Secondary Suspension)
- ⦚ Lateral Potentiometer (Secondary Suspension)
- • Vertical Potentiometer (Primary Suspension)
- ┃ Lateral Potentiometer (Primary Suspension)
- ⎯ Longitudinal Potentiometer (Primary Suspension)
- ▭ Bogie Yaw Potentiometer

Fig.64 - Plan of typical ride test instrumentation, bogie vehicle

Each transducer was connected by an individual instrumentation cable that was secured along the side of the vehicle underframe using cable ties, and running back to the test coach for connection to the instrumentation conditioning and recording equipment. Before commencing any test installation the test engineer in charge defined the detailed scope of the instrumentation to be fitted, and prepared a diagram to outline the type, quantity and location for each transducer to be installed. The test engineers then drew up the necessary request for equipment form which was presented to the Testing Section instrument stores, where the necessary test equipment was prepared ready for collection prior to installation to the vehicle. In order to measure the accelerations on axleboxes, bogie frames and vehicle underframes required to derive the acceptability of the running characteristics, then accelerometers of various ranges and types were used. Different ranges of transducer were selected based on where the measurements were to be taken. For example if a measurement was required at axlebox level then an accelerometer with a higher rating and measurement range was required compared to that of one installed on the vehicle underframe; this was because the magnitude of accelerations seen during on-track running were far greater at axlebox level. The general rule of thumb for use of accelerometers was:-

- Axlebox level = accelerometers with $^+/_-$ 100 g range,

- Bogie Frame or freight vehicle (with single stage suspension), mounted at underframe level = accelerometers with $^+/_-$ 15 g range,

- Passenger vehicle (with 2 stage suspension) mounted at underframe / floor level = accelerometers with $^+/_-$ 2 g range.

Fig.65 - Typical accelerometer and potentiometer installation on a wagon

In addition to the use of accelerometers, other transducer types such as linear measurement potentiometers were used to provide test engineers with information relating to the movements within the suspension of the vehicle whilst under test.

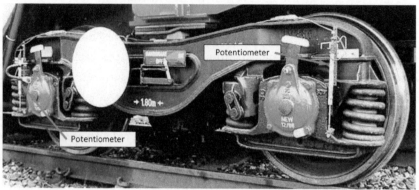

Fig.66 - Typical vertical suspension displacement potentiometer installation

As with the accelerometers, different sizes of potentiometer were used dependent on the location and magnitude of the measurements to be taken. For example the measurements of the primary vertical suspension movement of a bogie freight wagon normally required a transducer with a range of $^+/_-50$ mm, whereas the lateral suspension movement of the same vehicle only required a transducer with $^+/_-15$ mm range.

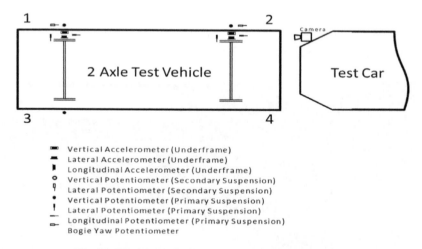

Fig.67 - Plan of typical ride test instrumentation, 2-axle vehicle

The positioning of transducers on the vehicle was extremely important not only in order to obtain credible measurements, but to be able to accurately establish how the vehicle and suspension was operating during the test. For example, with reference to the diagram in Fig.67 above, by mounting vertical suspension potentiometers at three corners of a 2-axle wagon, it was possible to compare the phase of the suspension vertical movement using the output of each of the three transducers to establish whether the vehicle was rolling, pitching or bouncing.

2-axle Vehicle Modes of Motion	Vertical Potentiometer Measurements "In Phase"	Vertical Potentiometer Measurements "Out of Phase"
Roll	1 and 2	1 and 3
Pitch	1 and 3	1 and 2
Bounce	1, 2 and 3	-

Table 7 - Example of establishing different vehicle modes using potentiometers

In a similar manner the dynamic behaviour of a bogie was established by installing potentiometers to measure vertical suspension movement at three of the corners of a bogie. When using acceleration measurements to assess the potential for derailment, only the off-loading direction of vertical accelerations needed to be considered, whereas both left and right lateral accelerations were analysed. It was also important therefore to understand the relationship between the direction of the vehicle body vertical accelerations and the forces on the track. The direction of actual vertical forces and vertical accelerations when a wheelset runs on an upward track irregularity the immediate effect was for the primary suspension springs to compress, this corresponds to the on-loading force at the wheelset. Conversely, when a wheelset runs on to a downward track irregularity (a track dip, or a dipped rail joint), the immediate effect was to extend the primary suspension springs; this corresponds to the off-loading force at the wheelset.

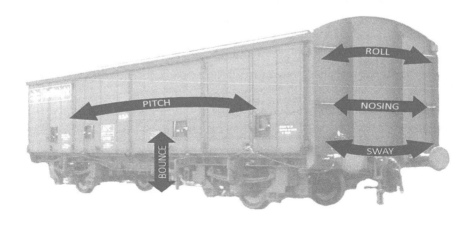

Fig.68 - Five basic modes of motion, 2-axle wagon

The mounting of accelerometers in the plane in which they were intended to measure had to be quite accurate to ensure that cross measurement of other planes was not picked up within the measurement. For example if an accelerometer mounted laterally on a vehicle underframe was not mounted at 90 degrees to the vertical, then any vertical acceleration was also recorded by the lateral transducer; in addition the accuracy of the lateral measurement was compromised. The differences in the design of the many vehicle types under test and the variety of transducers needed, meant that in most cases the installation of the temporary test instrumentation was very much tailored to suit each series of tests. When I joined the DM&EE testing section, there were a few common practices in place that had been developed over the years and adapted to suit the arduous environment on the outside of rail vehicles when under test; particularly on the underframe, bogies or axleboxes of freight vehicles. Basic rules and standards were adhered to as follows during the 1980s in preparations for dynamic test running:-

- All transducers, associated fixing brackets, cables etc. were secured such that they remained within the gauge envelope of the vehicle under test.

- All transducers and associated fixing brackets were secured remaining in-place when subjected to the g-forces seen during operational running, including shunting shocks. The British Rail standards available at the time from the structural engineers department provided the g-force level for which equipment mounted on rail vehicles was required to withstand during operation. Where necessary calculations were carried out to establish that a suitable design of fixing for transducers were used.

- All transducers and associated fixing brackets had a secondary retention strap for retention in case of failure of the primary fixing method in instances such as being struck by ballast, birds or other foreign objects during test operations.

- Where instrumentation cables had to be routed between adjacent vehicles or between parts of the vehicle that moved such as between the bogie and the underframe; then sufficient slack was maintained to cater for the extremes of movement whilst ensuring that cables could not protrude outside the gauge envelope of the vehicle or become entrapped in other moving parts.

The most common method of attaching test equipment such as accelerometers or potentiometers to a vehicle underframe or bogie frame was to use a thin metal bracket or plate bonded to the vehicle using a well-known brand of car body filler to which the transducer was then bolted using 5 mm diameter bolts and nuts with spring washers. Before the car body filler could be used to attach the instrument bracket, the surface of the vehicle where the bracket was to be attached, was prepared by removing the paint and roughing the surface to provide a good key for the car body filler to adhere to. Preparing the instrument bracket in the same manner, it was then just a case of applying the car body filler and holding in place for a short period until it had set; which was not normally too long, particularly on a warm day, unless you forgot to mix-in the hardener first. Other methods of attaching transducers included directly bolting them in place using apertures or holes where available on the vehicle, and by using proprietary engineering clamp brackets. The next stage was to connect the transducers up the instrumentation within the test coach via the cable plugboard mounted inside the leading end of each test coach. The instrumentation cables were normally made up from a 6-core screened PVC industrial type cable in varying lengths from 5 metres up to 50 metres, with die-cast Plessey type end connectors of Air Ministry origin. This type of connector was used commonly during the 1980s on DM&EE test equipment, including all transducers and amplifiers/signal conditioning units.

Fig.69 - Plessey 6 pin connection plug used for most test equipment and cables

The internal cables within the coach from the rear of the plugboard connected to the rear of the purpose built amplifier units which provided the electrical power to each transducer and the signal conditioning of the transducer outputs ready for recording. The set-up of each individual data channel was required to provide guidance for the test engineers undertaking the task, traceability of the test installation and important information to support the post-test analysis process. A typical list of the information recorded in a data sheet included the following:-

- Test Date and Project Description

- Test ID Channel Number

- Measurement Description

- Transducer Type and Range

- Transducer Asset Number

- Fixing Method for the Transducer

- Instrumentation Cable Number(s)

- Amplifier Rack Asset Number

- Amplifier Asset Number

- Nominal Sensitivity

- Calibration Value (Engineering Units)

- Inferred Calibration Switch Position

- Filter Setting (pre-Recording Filter)

- Calibration Output Voltage

- Tape Recorder Track Number

- Chart Recorder Channel Number

- Tape Recorder Asset Number

- Chart Recorder Asset Number

In the mid-1980s before the advent of computers, this information was hand written for each test, a time consuming task for the junior test engineer, considering that in the order of 20 data channels per test could be needed. The transducer and signal conditioning amplifier sensitivity and pre-recording filter settings were selected based on the test output requirement and the vehicle type. For a typical two axle leaf spring freight wagon ride test the following gives an example of the settings used:-

Measurement	Amplifier Sensitivity	Tape Recorder Input Filter Setting	Tape Recorder Input	UV Chart Recorder Input Filter Setting	UV Chart Recorder Sensitivity
Body mounted vertical acceleration	1 g = 1 volt	10 Hz Low Pass	1 g = 1 volt	6 Hz Low Pass	1 g = 2.5 cm
Body mounted lateral acceleration	0.5 g = 1 volt	10 Hz Low Pass	0.5 g = 1 volt	6 Hz Low Pass	1 g = 5 cm

Measurement	Amplifier Sensitivity	Tape Recorder Input Filter Setting	Tape Recorder Input	UV Chart Recorder Input Filter Setting	UV Chart Recorder Sensitivity
Primary Suspension Vertical Displacement	30 mm = 1 volt	Unfiltered	100 mm = 1 volt	Unfiltered	30 mm = 3 cm
Primary Suspension Lateral Displacement	20 mm = 1 volt	Unfiltered	50 mm = 1 volt	Unfiltered	20 mm = 2 cm

Table 8 - Table of typical measurements settings

Once the set-up of the instrumentation was logged and the amplifiers powered and warmed-up then the process of adjusting each amplifier unit with the required settings could begin. It was important to get the sensitivity settings correct for the specific test in order to capture the data required with sufficient resolution to be useful for on-board assessment on the chart recorders or for post-test analysis.

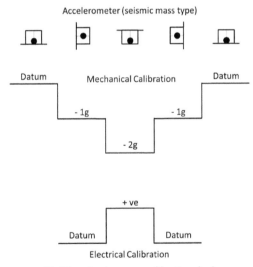

Fig. 70 - Accelerometer calibration checks

For example if the expected range of displacement of the suspension of a freight wagon was +/- 20mm then a nominal sensitivity of 30 mm = 1 volt was used to ensure ability to record greater than expected in the case of traversing a poor track feature which may cause a greater suspension movement than expected. If the sensitivity range was not selected correctly then it could lead to over-recording whereby the amplitude of the data signal output from the amplifier could exceed the input range of the data recorder. Once the electrical settings for each transducer were selected on the amplifier and the output balanced to read zero, a physical (or mechanical) calibration check was carried out by physically moving the transducer to generate a known output. For seismic mass type accelerometers this was a straightforward process of placing the accelerometer on a flat, level surface, then rotating the accelerometer through 90 and then 180 degrees whilst recording the output onto the chart paper trace recorder. The calibration checks for potentiometers could also be easily undertaken by moving the spindle of the transducer in and out by a known distance. To ensure that an accurate and consistent calibration check was applied, we manufactured stainless steel plates with 20 mm steps which were used to guide each potentiometer stepped calibration movement.

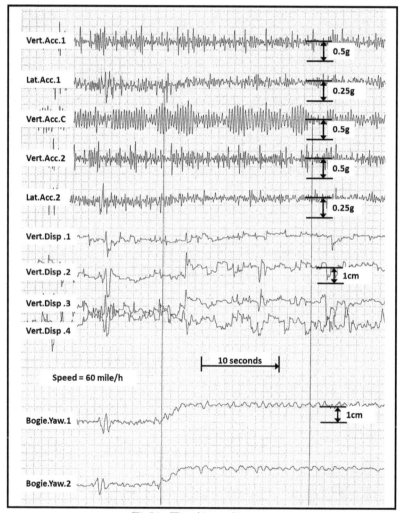

Fig.71 - Time history chart trace

The use of paper trace charts was vital during and also post-test for establishing the safe running characteristics of the vehicle under test and whether the suspension was performing as designed. The chart traces also provided a visual confirmation that all data channels were working correctly at all times during the test run. In the mid-1980s the majority of the chart recorders in use were manually operated SE 6300 type ultraviolet (UV) paper trace recorders; however from about 1990 onwards thermal paper recorders were becoming more common and could be controlled by computers. An example of a typical thermal paper chart trace is shown above in Fig.71. This particular trace was typical of a the output produced during a laden condition freight vehicle test run with a high friction content suspension such as a vehicle fitted with a GPS type bogie. The chart recorder paper feed speed was 5 mm/second which provided an adequate resolution to determine the frequency content and amplitude of the accelerations. Should the frequency of the traces increase then during the test run it was common practice for the Officer in Charge (OIC) to increase the speed of feeding of the paper in the chart recorder to provide an expanded time history trace of the suspension activity.

Time History Trace (Fig.71) Key:-

- Vert.Acc.1 = Vertical Body Acceleration above Bogie 1

- Lat.Acc.1 = Lateral Body Acceleration above Bogie 1

- Vert.Acc.C = Vertical Body Acceleration at the Centre Body

- Vert.Acc.2 = Vertical Body Acceleration above Bogie 2

- Lat.Acc.2 = Lateral Body Acceleration above Bogie 2

- Vert.Disp.1 = Vertical Primary Suspension Displacement at Wheel 1

- Vert.Disp.2 = Vertical Primary Suspension Displacement at Wheel 2

- Vert.Disp.3 = Vertical Primary Suspension Displacement at Wheel 3

- Vert.Disp.4 = Vertical Primary Suspension Displacement at Wheel 4

- Bogie.Yaw.1 = Bogie Yaw Displacement of Bogie 1

- Bogie.Yaw.2 = Bogie Yaw Displacement of Bogie 2

Some conclusions that can be easily drawn from this example time history trace include the following :-

- The vertical body accelerations measured above the bogies were well damped with no cyclic activity; the levels of acceleration at 60 mile/h were well below

0.4 g off-loading and the mean acceleration level was much lower therefore the suspension was behaving as expected and easily met the acceptance criteria.

- The lateral acceleration levels above the bogies were also well damped with no cyclic activity. The levels of acceleration were well below 0.25 g with the exception of a single incidence that was indicative of being caused by a track irregularity, although the suspension had coped with this well and was expected to easily meet the acceptance criteria.

- The vertical acceleration at the centre of the vehicle body was showing cyclic oscillation indicative that the underframe of the vehicle was flexing a little, however the level of acceleration was well below 0.5 g and because there was no influence noted on the body accelerations above the bogies then there was little cause for concern.

- The vertical suspension displacements confirmed the suspension was behaving as expected and damping out track imperfections well with no sign of any resonance.

- The bogie yaw displacements were showing very low levels of movement and no sign of oscillation at a level that indicated instability. The change in level part way through the trace was showing the bogie rotation whilst exiting a curved track section and onto a straight track section.

The OIC undertook just such an assessment of the performance of the vehicle under test at each constant test speed to establish whether it was safe to increment the speed further. The chart recorder operator was responsible for annotating the chart record as the paper fed out of the recorder, gathering the information to write on the charts from the commentary provided from the cab, the speed display in the test coach and looking out of the window to capture locations or specific conditions or infrastructure features that could affect the results in any way. As the on-board analysis systems were developed in the 1990s the annotation of chart records was assisted by automated input from the computer, however a chart operator still had to annotate traces with additional information. In the 1980s the primary method of recording data during on-track test runs was by magnetic tape recorders that used one-inch data tape reels. The 14-track versions of these large machines required two people to lift and were mounted within instrument racks in the test coaches for security and to limit vibration movement during operation. In the late 1980s, the newer 42-track versions of the one-inch tape recorders provided greater channel capacity. The one-inch magnetic tape reels were booked out and collected before each test from the DM&EE test section data analysis laboratory at the RTC Derby. The analysis laboratory ensured each registered data tape was clean and degaussed (to remove all existing test data from previous tests) before issue. Often for a larger series of tests, with more than one data recorder installed in the test coach and covering a few days away from Derby, then a large quantity of tapes was needed. Cleanliness of the tapes and the data recorders when in use was paramount. The tapes were shipped to and from the analysis laboratory in sealed protective cases and before each tape was installed the tape recording heads were cleaned with cotton buds soaked in an alcohol solution. For undertaking testing such as ride performance tests on a freight wagon the data recording speed was normally set at 1.875 inches per second (IPS) to provide sufficient bandwidth to record up to 100 Hz on all 14 data channels. Each tape being 4600 ft in length provided over 8 hours recording time on one tape. Other tests that required the capture of data at higher frequencies such as for structural assessment where data may need to be recorded at frequencies up to 1000Hz, then faster tape speeds were used; this meant that the data tapes required changing more frequently and more tapes were needed per test. During the regular freight ride testing a tape coding system was used to facilitate automated post-test analysis and an oscilloscope was used to monitor the data output real time from the tape recorders ensuring validity of the data, confirming that recording levels were not being exceeded and data was not affected by electrical interference. The nature of the ride testing meant that the data was collected during periods of running at constant speeds; the post-test analysis was undertaken by selecting these constant speed sections based on a code signal created by the test engineer, which was recorded onto one of the magnetic tape data tracks. This signal coding, known as Z-P coding was created by a specially designed and built code generator unit installed in the table top rack beside the main instrument rack in Test Car 1 alongside the oscilloscope and Time, Speed and Distance conditioning unit.

The Z-P code unit provided the following functions:-

- Z or Zero code, used to denote the start of a constant speed test section.

- P or Print code, used to denote the end of a constant speed test section.

- D or Disable code, used to pause the constant speed test section should the train speed fall outside the $+/- 2.5$ mile/h boundaries.

- E or Enable code, used to un-pause the test section.

It was the responsibility of the test engineer in charge of the tape recorder to operate the coding unit throughout the test run such that the data on the tape was correctly coded to allow automated Peak Count Zero Crossing (PCZC) analysis when the tapes were replayed back in the analysis laboratory. By the late 1980s the technology for data tape recording had moved away from the bulky and fragile one-inch data tape recorders to more compact and robust VHS tape based recorders. These tape recorders still required the operator to maintain the cleanliness of the tapes and the tape recording heads still needed to be cleaned with cotton buds soaked in an alcohol solution. The Freight Group of the testing section were the last of the DM&EE test groups to move to using the new TEAC sourced 21-channel VHS tape data recorders in about 1990; at the time when the development of the on-board analysis system for freight vehicle ride testing was well under way for use in Test Car 1. The first generation on-board analysis system often referred to as the OBA, used for ride acceptance testing was developed by me in the early 1990s. A Hewlett Packard (HP) Basic 9000/320 computer system was used to process, store and print the analysed results for ride acceptance. The standard instrumentation set-up was maintained with accelerometers connected to conditioning amplifiers providing data signals that were recorded onto VHS tapes and a thermal chart recorder. In parallel, 6 Hz filtered acceleration data signals were fed into a Stress Engineering Services six channel data cycle counting unit which had EPROMs (Erasable programmable read-only memory) configured to undertake the PCZC cycle counting on each data channel. The methodology of selecting the constant speed test sections using the Z-P coding was also retained, but now controlling the operation of the data cycle counting unit through the computer and automatically downloading and the analysed data from each test section as the test run progressed. I developed the Basic code program to interface between the analyser and the computer which included functionality for printing the test log sheets semi-automatically and annotating chart records on the new thermal paper recorders. This used a two digit code system as shown in Table 9, whereby the OBA operator entered codes into the keypad as the test progressed, from a pre-defined list covering test route locations and relevant information such as curves and track features etc.

OBA Codes		OBA Codes	
01	Left Hand Curve	05	Facing Point
02	Right Hand Curve	06	Crossover
03	Straight Track	07	Welded Track
04	Trailing Point	08	Jointed Track

OBA Codes			OBA Codes	
09	Accelerating		26	Wrong Line Working
10	Decelerating		27	In Sidings
11	Braking		28	Stellites
12	Temp. Speed Restriction		29	Moving Off
13	Main Line		30	Loss of Power
14	Slow Line		31	Locomotive Change
15	Dry Rail		32	Level Crossing
16	Wet Rail		33	Foot Crossing
17	Train stopped		34	Tractor Crossing
18	Signal Check		35	Passing Train
19	Tape Rec. Off/On		36	(spare)
20	Tape Calibrations		37	Cant Deficiency
21	Tape Datums		38	Cant Excess
22	Loco. Run-Round		39	Tunnel
23	Train Turned		40	Rising Gradient
24	Pumping Sleepers			
25	Wet Track Spots			

Table 9 - Table of two-digit codes for the OBA system

After completion of the outward leg of a test run, there was usually some time whilst the locomotive run-round was happening for the OBA operator to produce test analysis from the outward run and output using selectable functions to choose which constant speed tests to include into the Freight Acceptance Curve plots. A Hewlett Packard type 7475A pen plotter was used to print out the analysis plots on-board the train which provided the Officer in Charge of the test with confirmation of the acceptability of the vehicle ride, and evidence of how much valid data had been collected on the outward leg of the test. This enabled the constant speed bands during the return test run to be selected and operated accordingly to ensure population of all the speed band data sets with the required distance of tests data ready for full analysis shortly after the return to the RTC.

Having covered what we were going to test and how we were going to record, let us now take a look at the test routes that were used for on-track ride testing. In the main, the freight and on-track plant machine ride testing was carried out between the Derby RTC and various points on the route via Melton Mowbray, Manton, Corby, Kettering to Bedford. The test runs normally terminated on the connecting line between Bedford Midland and Bedford St Johns stations where the locomotive ran-round. As the speeds of new freight rolling stock increased during the 1980s the test runs were often extended past Bedford in order to provide for more representative operations, continuing onwards towards London on the slow lines terminating at the Recess Sidings at Cricklewood.

Fig.72 - Freight Acceptance test route, 1980s

The length of the test run was normally planned based on the maximum speed of the vehicle under test, therefore vehicles with a higher designed operating speed (say 75 mile/h) underwent a test run all the way to Cricklewood sidings, whereas a vehicle with a slower designed operating speed (say 45 mile/h) was normally only tested as far as Leicester or Manton Junction. In all cases it was essential that sufficient valid test data was obtained from a test run to fully assess the behaviour of a vehicle as representative of behaviour in service. It was also essential to assess the potential for wheelset instability (hunting), which was affected by speed and by the coefficient of friction between wheel and rail. This coefficient was reduced when the rails were greasy or damp, therefore to mitigate the risk of not establishing whether hunting was prevalent, especially when running on poor rail conditions, it was normal practice to increase the maximum test speed to 10 mile/h over the maximum designed operating speed. For example when rail conditions were any less than totally dry, or when there was any signs at all that stability performance may not have been fully established, a vehicle with a designed maximum operating speed of 60 mile/h was tested, and was expected to meet the Freight Acceptance criteria at speeds up to 70 mile/h. The test route incorporated a large selection of representative infrastructure features, different grades of line with a range of line speeds and different rail types. By the mid-1980s the majority of the Midland Main Line route fast lines were installed with continuously welded rail (CWR), whereas the goods lines and large sections of the route section between Syston Junction via Melton Mowbray, Oakham, Manton and Corby to Kettering still had jointed rail and in some cases older style bullhead type jointed rails. The track type and track quality was particularly important for the assessment of vehicle suspension systems which may be affected adversely by the cyclic inputs seen when traversing track features and in particular jointed track. The track condition was regularly assessed by reviewing infrastructure data recorded using the High Speed Track Recording Coach (HSTRC) RDB999550, to ensure that the track quality over which vehicles were tested remained representative. In addition to this, and to help the test engineers evaluate any discrete track features during a test run, a 2-axle covered ZXA van number ADC201055, was included in the test train formation for the purpose of providing ride performance data from a known 'monitor' vehicle.

Fig.73 - ZXA Taperlite suspension [Mike Fraser]

This vehicle was used for this purpose for at least 17 years and being fitted with a Taperlite hydraulic damped suspension provided a reliable consistent method of assessing discrete features and establishing whether the test route track quality was remaining representative. Should the ZXA vertical or lateral accelerations recorded when traversing a discrete track feature be considered out of character for a particular section of track, this indicated that the track feature traversed was of an unacceptable level. At the discretion of the Officer in Charge (OIC) the data section recorded for the vehicle under test, whilst running over the particular track feature was assessed and justifiably removed from the data analysis providing it was considered unrepresentative of the overall vehicle performance during the test. In later years following introduction of advanced computer based data collection and analysis, the test data recorded that was used in the analysis and assessment for acceptance of vehicles was automatically selected from that recorded when operating over track sections that met the current $1/8^{th}$ mile Standard Deviation (SD) criteria for rail top or alignment irregularity. The strategy to reduce the levels of jointed track on the network, particularly on primary routes was implemented during track renewal maintenance of the infrastructure, and from the early 1990s we saw a marked reduction in jointed track on the Midland Main Line test route particularly over the section between Syston and Manton Junction. The general reduction in freight traffic also lead to the removal of some of the goods lines, in notably the section between Kettering South Junction and Harrowden Junction (north of Wellingborough), which saw the goods lines removed completely; and the sections between Corby and Kettering and also the goods lines between Wellingborough and Sharnbrook Junction which were reduced to a single line. It is ironic that within a few years after removal, the freight traffic had picked up again sufficiently to warrant the re-installation of the goods line between Kettering South Junction and Harrowden Junction, albeit only with a single line. As a result of the reduction in jointed track and the inflexibility of test train operation over the Midland Main Line route around Kettering, the test team started to investigate alternative routes.

By late 1995 most on-track ride tests had moved away from the Midland Main Line and were normally being carried out over the route between Derby, Crewe, Warrington and where higher speeds data collection or longer test runs were required, to Carnforth. Although most new designs of rolling stock exhibited riding properties that were acceptable during their first test runs, in some cases the performance of the vehicle suspension did fail to meet the criteria, and as such the maximum speed was subsequently limited. In such instances the test run was normally cut short, for example during a planned test run to Cricklewood, the riding properties of a vehicle designed to operate at 75 mile/h were assessed by the Officer in Charge (OIC) on-board to be unacceptable at speeds of 55 mile/h and above; therefore the test run was terminated at Kettering and returned to the RTC Derby at a maximum speed of 50 mile/h. A typical freight vehicle test train was formed with the vehicle under test and the trailing/north end of the train when leaving Derby as follows:-

Fig.74 - Outward test run formation - southbound

The normal practice was for the outward test run to leave Derby with the vehicle under test being loose coupled to the trailing observation end of the Test Car. The term loose coupled meaning that the screw couplings between the vehicle under test and the test coach were not screwed up tight leaving a gap between the buffers.

Fig.75 - Return test run formation - northbound

The reason for loose coupling was to ensure representative running conditions and the vehicle under test was not artificially stabilised by tight coupling to the adjacent test coach or locomotive. The locomotive was run-round to the other end of the train at the outward destination, and loose coupled to the vehicle under test for the return trip back the Derby. During ride tests an intercom system was provided between the test coach and the locomotive cab which enabled the Officer in Charge (OIC) to communicate the required operating speeds to the traincrew. It was also used by a member of the test team riding in the locomotive to provide a running commentary back to the test coach about the test route. This information was broadcast over loudspeakers within the test coach instrumentation saloon area and also recorded onto the magnetic tapes along with the test data. When the tapes were replayed for post-test analysis, the live commentary could be used to help understand locations and running conditions when evaluating the results from the tests; and provided key information that the chart recorder operator hand-wrote on the paper alongside the test results traces as they emerged from the recorder. An example of the sort of information that was provided in a commentary from the cab included:-

▪ track type, welded or jointed rail

▪ switches and crossings

▪ mileposts (to confirm location)

▪ stations

- overhead line features (for pantograph testing)
- track irregularity features such as wet spots or alignment
- curves (right hand or left hand)
- traction power demand
- braking activity
- tunnels
- speed restrictions

During most ride test runs, a traction inspector as well as the driver was on-board, therefore because most locomotive cabs only had two seats, the member of test staff riding in the cab did not have anywhere to sit. In order to provide a bit of comfort during a 3+ hour test run, a stool was normally used to perch on in the cab; these were a specially made type of stool that was the right height to allow visibility out of a locomotive cab window, and easily fitted through a cab door opening; although for some locomotive types such as in Class 37 cabs, where the windows were much higher in relation to the floor level the stools were not much use. Thankfully Class 37 locomotives were rarely provided for the ride test runs.

A typical freight wagon ride test with Test Car 1 (TC1) running from the RTC Derby to Bedford and back operated along a well tried and tested schedule which normally happened something like this:-

07:30   Arrive
Engineers arrived at the RTC and boarded TC1 which was normally still parked within the EDU workshop. Power supplied by the workshop shore supply was switched on, as was the test instrumentation in order to get it warmed up.

07:45   Instrument Checks
Once the instrumentation was warmed up, the process of balancing each signal conditioning amplifier data channel began, followed by checking of the calibration settings for each data channel. The magnetic tape recorder tape heads were cleaned and the tape reel loaded and the chart recorders were loaded with new UV paper rolls.

08:15   Calibrations
The tape and chart recorders were then started and a series of calibration checks recorded and inspected to ensure all was in order. The recorders and test equipment was then switched off, the shore supply cable disconnected from the workshop wall socket, coiled up, and stored in the workshop at the rear of TC1.

The Officer in Charge (OIC) was normally on-board TC1 by this point to ensure all was in order.

08:30 Shunt Move

The resident RTC Class 08 shunt locomotive arrived and was coupled to the test train to draw the formation out of the workshop into the RTC Way & Works sidings ready for the arrival of the main line locomotive. If fuel was required for TC1s generator, then the shunt move proceeded to the loop road at the north end of the RTC sidings where the fuel wagon was located. The shunting staff assisted with the fuelling whilst the test engineers checked the oil and water levels in TC1s generator and the generator unit started once the fuelling process had finished. The test instrumentation was then switched back on, as well as the kettle!

08:45 Test Locomotive Arrival

Following arrival of the test train locomotive, the RTC yard staff coupled the locomotive to the south end of the test train and carried out the statutory brake test, whilst the test engineers fitted the intercom system cable between TC1 and the locomotive leading cab using cable ties to secure it along the side of the locomotive. Before departure an intercom system function check was carried out and one of the test team joined the driver and traction inspector in the locomotive cab to provide commentary during the test run; normally with cups of tea in hand.

09:00 Departure time

Although when I first joined the test section it was accepted practice to propel the test trains in or out of the RTC sidings, by the late 1980s this practice was stopped and a shunt locomotive was used to haul the test trains out of the sidings. After checking with the Officer in Charge in TC1 that everyone was on-board and ready, the Class 08 shunt locomotive, which was still attached to the north end of the test train, hauled the whole formation out of the RTC sidings. Dependent on which line the train was departing on, this was either into Derby station or onto the loop line on the curve alongside the locomotive stabling sidings, whereby the Class 08 locomotive was detached and the test train awaited the signal to be cleared for departure southwards along the Midland Main line.

09:15 Test Run

The booked path of the test train normally followed a local passenger service out of Derby. Once we had the clear signal, the Officer in Charge (OIC) requested (via the intercom system) for the train to draw forward onto the main line straight track section alongside milepost 128 outside the RTC. Here the tape data recorders were started, a final instrument balance check was carried out and the distance counter set to 128 as the test car passed the milepost. Once these checks were completed which normally took less than a minute, the OIC instructed to accelerate the train up to an initial test speed of 30 or 40 mile/h, and the test engineers started the paper trace recorders. The OIC monitored the records on the paper trace recorders to establish that the riding properties of the vehicle under test were acceptable; the observation windows provided good visibility to allow the vehicle under test to be observed continuously. Once the OIC was happy that sufficient test data had been collected at the initial test speed, the instruction was given to increment the speed up by 5 mile/h, following the same process again, each time assessing the riding properties of the vehicle were acceptable before increasing the speed to gather more data. The rail type on the main line heading out of Derby was welded rail, whereas the route after passing Syston Junction on the Melton Mowbray line and via Corby during the 1980s included long sections of jointed type rail. Some vehicle suspension systems were affected adversely by the cyclic inputs seen when traversing jointed type rail, therefore the process of data collection and assessment of the riding properties began at a low speed on the jointed track sections, incrementing up the speed in the same manner to record evidence before going faster. During the test run it was common practice for the OIC to increase the speed of feeding of the paper in the chart recorder, to provide an expanded time history trace of suspension activity where necessary to see greater detail.

11:30   End of outward run

Depending on the performance of the vehicle under test, and the speeds attained during the test run the arrival time at Bedford varied. Normal practice during the 1980s was to move the test train though Bedford Midland and onto the loop line towards Bedford St. Johns station near the river bridge.

12:30   Run-round

The intercom system cable was disconnected and the locomotive detached to run-round; the whole run-round process took between 30 minutes and 1 hour depending on whether there was any attention necessary to the test instrumentation or inspection of the vehicle under test. Following re-connection of the intercom cable between the test train and the locomotive and swapping the stool into the other cab, it was normal for a different member of the team to head up front to provide the commentary for the return trip, again with cups of tea in hand for the traincrew.

Fig.76 - The observation end view from Test Car 1, July 1987

12:45   Return Test Run

The normal route for the return trip was via Leicester, probably the busiest station en-route at which we passed at slow speed, normally through platform 2, with many people on the platform taking an interest in the rather strange train formations and engineers wearing orange vests peering at them from behind rolls of chart paper, usually with a mug of tea in hand. The return trips were normally with the observation end of TC1 leading, and particularly when testing low height wagons the view forward was great. On days when the weather was rather inclement, the observation end was a draughty place and we were thankful for the 3 kW fan heater. In later years the installation of window wipers on the left and right observation end windows helped the view out when it was raining heavily.

14:30   Back at Derby

On arrival back at Derby the route to get us back into the RTC sidings depended on whether the RTC shunt locomotive and crew were ready for us. Providing they were ready and waiting at the RTC sidings signal, and there was sufficient headway before the next train following us, the test train was stopped with the locomotive just outside the Derby signal box on the down main line, and the Class 08 shunt locomotive then attached to the rear of the test train to haul us directly back into the RTC sidings. If the shunt crew were not ready, as often they were not, then the test train had to proceed into a vacant platform or onto the goods lines at Derby station and await the shunt locomotive arrival to drag us back into the sidings. The appearance of a test train in Derby station was not unusual due to the proximity of the RTC, however when waiting in a platform there was always interest from passengers, and also from station staff who were mainly concerned with how quickly we were going to get out of the way of passenger services. On the rare occasion when the shunt locomotive had failed or the RTC crew were unavailable then the test train locomotive ran-round the train in the platform and hauled it back to the RTC sidings.

14:45   Packing up

Once back into the sidings the intercom cables were removed from the test train locomotive, and not forgetting to collect the stool and teacups from the cab, the locomotive departed the RTC sidings to head back to 4-Shed, Toton or Crewe depot. The Class 08 shunt locomotive then moved the test train formation back into the EDU workshop, providing space was available, or onto one of the siding roads at the south end of the RTC site. The shunt movements normally provided time for the OBA operator to select and output the analysis from the whole test run. Whilst this was printing out, we undertook final instrument checks, recording of post-test electrical calibrations onto the chart and data tapes, packing and labelling the data tapes and chart records, and collating all log sheets and analysis information from the days test. Once the train was stabled securely and analysis plots finished the test car instrumentation was shut down and we returned all information back to the Testing Section office.

Before the freight acceptance on-board analysis system was developed, and where more specialised data analysis was required the data tapes and relevant test logs were submitted by the test engineers to the DM&EE data analysis laboratory, along with a detailed request for analysis form outlining the specific requirements for post test data analysis. In the mid-1980s when I joined the test section most data analysis was either undertaken by manual assessment of chart traces. or by using early Hewlett Packard (HP) technical desktop computers such as HP type 9845 machines, running basic code programs to interface with hardware analysers such as Stress Engineering Services cycle counting units or HP 3585 Spectrum Analysers.

By the late 1980s the analysis laboratory had standardised on the use of HP 9000/300 series computer systems which had moved on to use industry-standard HP-UX based data processing packages and instrument controllers, with bespoke compiled 'C' and Pascal subroutines for specific testing analysis requirements. These systems allowed the raw analogue data recorded on the tapes during the on-track testing to be digitised and the sampled data stored on hard disc drives for processing. Disc drive capacity developed very quickly from 40 MB drives in about 1987 to around 500 MB in 1989. The increase in the discs storage space and the speed of the processors in the computers allowed for greater data sampling rates, therefore permitting the digitisation process to be carried out for longer test sections at faster than real-time. The speed at which this could be realised was dependent on the maximum frequency to which the data was required to be evaluated. For example to assess the resonance or body-mode from vehicle suspension data, a 10 Hz low-pass filter was applied because frequencies within the data signal above 10Hz were not of interest. Data signals therefore needed to be digitally sampled at a rate sufficient to capture up to 10Hz. The sampling process however required over-sampling of the data to ensure the full signal was captured and to prevent what is called 'aliasing', which is the incorrect representation of an analogue signal in digital form. For example, analogue data signals must be sampled at a sufficient enough rate to accurately reconstruct the signal. If the data sampling rate is too low compared to the signal frequency content, then an incorrect representation of the signal may be generated. During digitisation a sample rate of at least twice the maximum data signal frequency must be used to correctly reconstruct the data signal with respect to frequency content; however to correctly reproduce the amplitude of the data signal, a ratio of ten times was normally used.

An example of the effects of under sampling can be seen in Fig.77, with Trace A showing a sample rate of four times the signal frequency therefore sufficient to re-reproduce the frequency content of the signal, and Trace B showing a sample rate too low compared to the increased signal frequency therefore digitally creating an incorrect representation of the signal; this is called aliasing. In most cases for freight vehicle assessment the requirement for data analysis was only 10 Hz, although the actual signal recorded on the data tape would have contained frequencies significantly higher, therefore an anti-aliasing filter was applied to the data signal between the analysis laboratory tape data player output and the digitiser input.

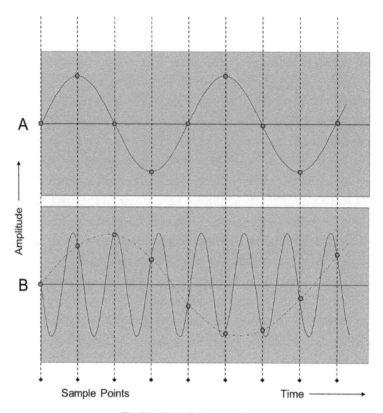

Fig.77 - Digital data sampling

Another benefit of applying pre-filtering was such that the maximum frequency content of the data signal was reduced to a level much closer to that for which analysis was required, therefore allowing the data sampling frequency to also be reduced to a much lower level without the risk of aliasing occurring, reducing the digital data file size and maximising the use of the computer data disk space. The majority of the freight vehicle ride tests did not involve additional data analysis over and above that produced from the OBA system; however when vehicles failed to meet the criteria further investigations were instigated which normally meant more work for the analysis laboratory.

In order to give some idea of the difference in ride performance between acceptable and unacceptable vehicles, the following four Figures show examples of Peak Count Zero Crossing cumulative acceleration analysis plotted against the limiting Freight Acceptance Curves for both vertical and lateral acceleration records from freight wagons.

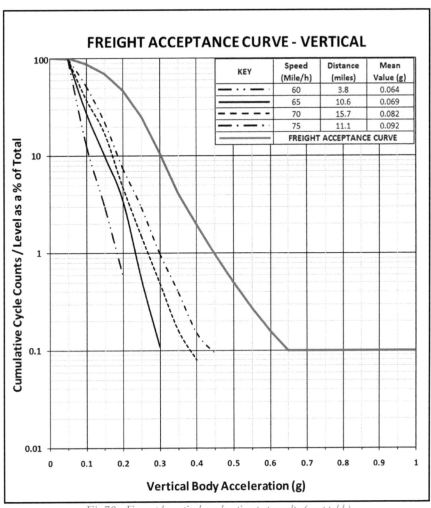

Fig.78 - Example vertical acceleration test results (acceptable)

The plot shown above in Fig.78 provides an example of the vertical performance of a good riding freight wagon operating at speeds up to 75 mile/h. Acceleration levels were well below 0.5 g, with no evidence of an inability to cope with any vertical track irregularities such as dipped rail joints.

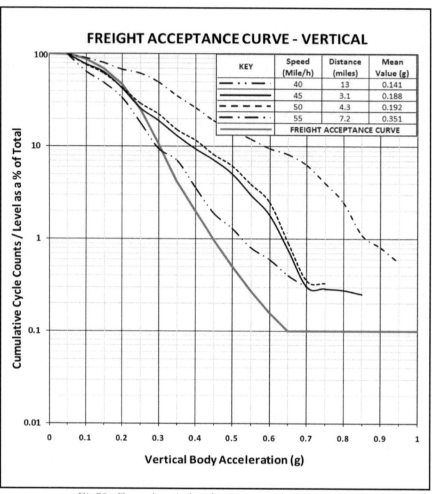

Fig.79 - Example vertical acceleration test results (unacceptable)

Fig.79 shows the vertical performance of a freight wagon running on jointed track that was deficient in damping, causing uncontrolled high acceleration levels, even at speeds below 50 mile/h. Scrutiny of the chart traces provided guidance to whether the vehicle was pitching or bouncing by comparing the acceleration traces at each end of the vehicle. The frequency content of the accelerations was also assessed to establish whether the predominant suspension frequency aligned with the rail joint passing frequency causing a resonance.

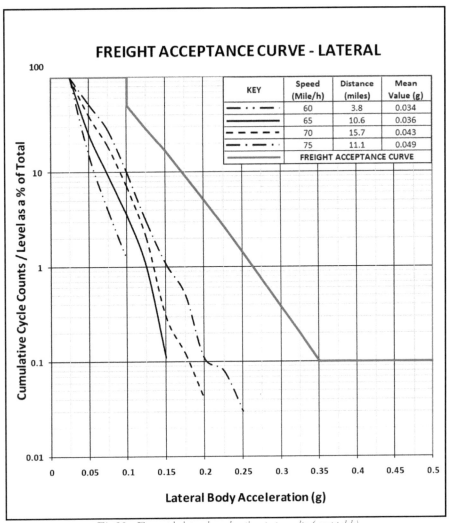

Fig.80 - Example lateral acceleration test results (acceptable)

Fig.80 provides an example of the lateral performance of a good riding freight wagon operating at speeds up to 75 mile/h; lateral acceleration levels were well below the limiting acceptance curve, there were no signs of instability and the vehicle coped very well with any track alignment irregularities or points & crossings.

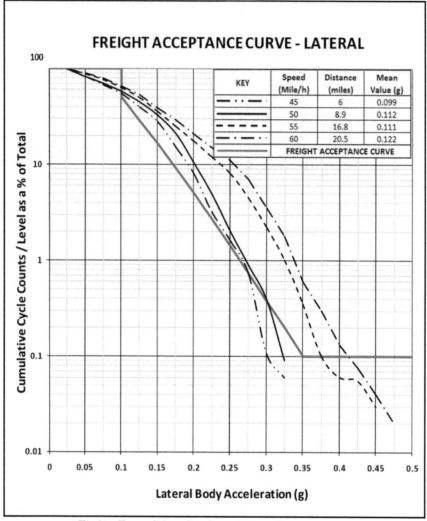

Fig.81 - Example lateral acceleration test results (unacceptable)

Fig.81 shows the lateral riding properties of a 2-axle freight wagon running on welded track with dry rails that was susceptible to uncontrolled hunting at speeds above 40 mile/h. This sort of lateral behaviour could have been caused by excessively worn wheel profiles, the incorrect wheel profile being used, or insufficient lateral suspension clearances or stiffness properties. Vehicle body roll, or sway could also have been prevalent when the vehicle was hunting, therefore the chart traces were scrutinised to establish any predominant body modes.

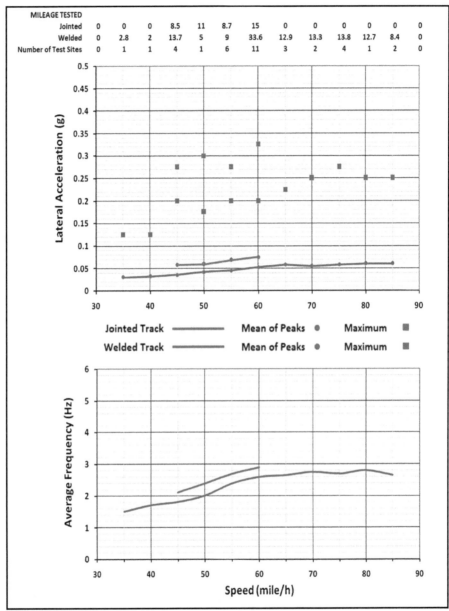

MILEAGE TESTED													
Jointed	0	0	0	8.5	11	8.7	15	0	0	0	0	0	0
Welded	0	2.8	2	13.7	5	9	33.6	12.9	13.3	13.8	12.7	8.4	0
Number of Test Sites	0	1	1	4	1	6	11	3	2	4	1	2	0

Fig.82 - Example lateral ride characteristics summary

The example summary analysis shows in Fig.82 the mean and maximum of the peak accelerations at each test speed, along with the average frequency of the lateral accelerations. The data analysis was split to show the jointed track and welded track characteristics separately because it was important to establish any resonance behaviour due to the jointed track cyclic inputs. Other the types of data analysis techniques undertaken within software include statistics of general time-based and frequency-based analysis of data signals, to specialist methods such as the assessment of the fatigue life of a vehicle structures or the equipment fitted onto a railway vehicle. Frequency based analysis was used to determine the predominant frequency of an acceleration data signal from a vehicle suspension in operation, or to highlight a resonance. Power Spectral Density (PSD) analysis looked deeper at the frequency analysis of a signal describing the power present in the signal as a function of the frequency.

An example of a PSD analysis of a vertical acceleration signal, recorded on cab floor of a locomotive, is shown below in Fig.83.

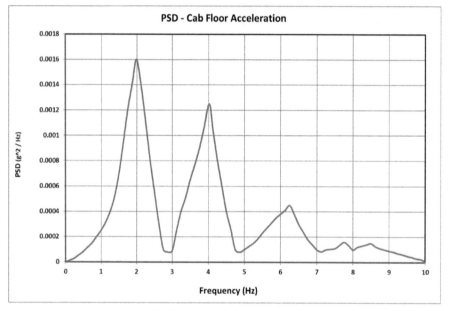

Fig.83 - Example of PSD analysis

Further development came during the mid 1990s with the increased capability of Windows based computer systems that significantly increased the range of software analysis tools available, and subsequently the speed in which analysis could be undertaken on large data files. The software programmes used were generally off-the-shelf products, however in some cases the Test Section Development team worked with an analysis product supplier to develop specific analysis functionality within the software to meet requirements relating to the rail industry. After data analysis was completed, the results were collated along with details of the tests into a formal report for the customer. These reports varied from a one page summary of findings to a fully detailed report containing photos, graph plots of the data analysis, diagrams, and a comprehensive write-up of the tests undertaken. As technology improved then so did the ease of production of these reports. In the mid-1980s many reports were hand written or dictated to the DM&EE typing pool staff, who prepared the text part of the report; the diagrams and analysis graphs were hand drawn where necessary using Letraset text transfers to improve the quality of any diagram text and headings. With the advent of word processors, the test engineers prepared the reports in full and attached the hard copy data analysis graph plots produced by the analysis laboratory; this continued until the evolution of Windows based computers which provided a fully integrated package for data analysis and reporting.

During my time at the DM&EE test section I prepared a large number of these technical reports covering tests I undertook, a summary of which is provided in the appendices at the back of this book.

The following part of this chapter looks at the experiences from a few of the many freight vehicle ride tests that I worked on, including various new bogie and 2-axle wagons, old wagons with new suspension or new bogies, vehicles modified to improve the ride, and vehicles for which the test runs provided a development platform to verify various design options.

Fig.84 - Prototype Prologie low floor container wagon, June 1985

One of the first full ride test projects that I was involved in after joining the test section was the prototype of a new ultra-low-floor bogie container flat wagon designed jointly by British Rail & Talbot and built by Procor for carrying 9ft 6in containers. The wagon designated TOPS code PFA, numbered PR95300 and referred to as a Prologie wagon, arrived at the RTC for a full suite of static and dynamic ride and braking tests in April 1985. The design incorporated small diameter wheels and running on aluminium in-board axlebox bogies allowed the floor height to be low enough to fit the taller height 9' 6" freight containers within the British Rail loading gauge. The normal static wheel weighing, torsional stiffness ($\Delta Q/Q$) and bogie rotation tests were completed followed by coupling to Test Car 1 ready for instrument fitting. This proved quite a challenge with the low underframe height and compact nature of the bogie and suspension arrangement. Following test preparations the first test run was operated on the 15th April from Derby with the wagon in its tare condition. Joining the test coach for my first Test Car 1 ride test, the coach was quite full with a number of visitors from the wagon manufacturer and the DM&EE Freight Design team; in addition to myself (in training) there were the normal four members of test section staff, being the Officer in Charge, a tape recorder operator, a UV chart recorder operator and the person who rode in the locomotive cab. Part of the assessment for the new Prologie vehicle involved a number of the runs with the small wheeled bogies passing over crossovers, which were undertaken at the Old Dalby test line. Opportunity was taken to record the ride and stability performance of the wagon on the movements from Derby to Old Dalby which was via Syston Junction and Melton Mowbray. The wagon was loaded with test weight containers at the RTC Derby and a further test conducted on the 25th April, again to the Old Dalby test line. Unfortunately the lateral stability performance of the new wagon was not as expected, and modifications were necessary that included the fitting of bogie yaw control dampers.

Fig.85 - Prologie wagon during test run at Syston Junction, June 1985

Retesting of the Prologie vehicle was carried out firstly with repeat bogie rotation tests, then on the 17th and 20th June 1985, tare and laden ride tests, on both occasions running from Derby to Kettering via Leicester and return via Manton Junction including a number of operational movements across the Glendon South Junction crossovers. It was during this series of tests that I had my first opportunity of providing the cab commentary during freight wagon test runs; learning very quickly the relevant features of the test route from Derby, supported by the traction inspectors on-board, who were always keen to share their many years of railway and route knowledge with test engineers. In the locomotive cab there was little indication of how the test was progressing other than the occasional comments from the OIC through the intercom system providing instruction to increase the train running speed or maintain it at the current speed because more data was required to confirm the ride of the vehicle under test. The operations for each test train path were pre-prepared, including departure, passing and arrival timings for key locations along the route, and issued in a Special Operating Notice (or SN) by the regional control beforehand. It was important for the test run to try and stick to the SN timings so that we could achieve the correct paths throughout the whole run, however being a test train, we were often held at various points along the route primarily to allow late running service trains the priority of the line. Whilst this was sometimes frustrating, it did on occasion provide opportunity to inspect the vehicle under test and rectify any minor instrumentation issues such as a faulty transducer or a loose cable connection. If it was absolutely necessary to access to non-cess side of the test train, then the traction inspector got in touch with the signalman to arrange a suitable slot in between services to stop other trains, thus allowing us to safely access the other side of the train. The test runs on the Prologie wagon also highlighted a number of issues with the bogie frame structure that were eventually overcome, and the vehicle did enter into service for a couple of years, although I do not believe any further vehicles were built to this design.

A number of older generation vehicles were assessed during the mid-1980s for effectiveness of suspension modifications applied in an endeavour to increase the operating speeds with a view to fitting air brakes enabling operation in fully fitted trains. Five Flatrol type 'Well' wagons, each having different types of suspensions springs fitted, were submitted for test to the RTC Derby between December 1985 and December 1986.

	DB900029	DB900030	DB900010	DB9000122	DB998014
Year Built	1956	1956	1952	1959	1959
Type	Flatrol SB	Flatrol SB	Flatrol MVV	Flatrol EAC	Flatrol WW
TOPS Code	ZVQ	ZVQ	ZVR	ZBW	ZVR
Wheelbase	26'2"	26'2"	24'0"	24'0"	25'0"
Spring Type	Coil	Volute	9 plate Leaf	8 plate Leaf	7 plate Leaf
Tare weight	15t 3cwt	15t 3cwt	10t 6cwt	15t 2cwt	14t 10cwt
Test Date (Tare)	16 Dec 85	7 Mar 86	12 Mar 86	11 Dec 86	11 Dec 86
Acceptable Speed	Not acceptable	25 mile/h	35 mile/h	35 mile/h	35 mile/h
Laden weight	34t 4cwt	34t 10cwt	29t 3cwt	33t 15cwt	33t 8cwt
Test Date (Laden)	12 Dec 85	12 Mar 86	7 Mar 86	18 Dec 86	18 Dec 86
Acceptable Speed	Not acceptable	25 mile/h	35 mile/h	35 mile/h	35 mile/h

Table 10 - Flatrol wagons, summary of vehicle details and tests

Torsional stiffness ($\Delta Q/Q$) tests were conducted and measurements of the wheel profile condition taken first before being coupled to Test Car 1 and applying simplified instrumentation to measure body accelerations and suspension displacements at three corners of each wagon.

Fig.86 - Flatrol wagon awaiting test at the RTC, 1986

The test runs in each case were relatively short, mainly due to the very poor riding properties of the wagons. In particular the test run with DB900029 only reached Spondon, a few miles from Derby before being terminated and the test train returned to the RTC at a maximum speed of 25 mile/h. Not much success was achieved in increasing operating speeds by making changes to the suspension of these Flatrol wagons, therefore very few were subsequently fitted with air brakes and retained for main line use.

Two of the early 1970s built FGA type freightliner container wagons, fitted with the original Ride Control type two metre wheelbase bogies, were submitted for test in late 1986 as part of investigations into improvements to the vertical riding properties. Wagon 601135 had standard friction damping material whilst wagon 601239 was fitted with a manganese friction damping material to enable a comparison of the riding properties to be undertaken. The empty wagons were delivered to the RTC where static torsional stiffness ($\Delta Q/Q$) tests were carried out before coupling to Test Car 1. Test instrumentation was installed to measure the riding properties, including underframe and bogie mounted accelerometers, and suspension potentiometers. The first test run, hauled by locomotive 47357 ran via Beeston Freightliner Terminal in order to load the wagons with test load containers. Following loading on the 19th November, we continued via Manton, Corby, Kettering and then on to Cricklewood, running at speeds up to 85 mile/h where the line speed permitted south of Bedford. The locomotive then ran-round in the Cricklewood recess sidings, re-coupled and the train returned direct via Leicester. The second test run on the 20th November also ran via Beeston in order to swap the loaded containers onto the other wagon, then continuing in the same manner as the previous days tests to Cricklewood before returning to Derby.

Fig.87 - Loading containers at Beeston Freightliner Terminal, November 1986

Following completion of the ride tests, the analysis undertaken of the ride test data showed the level of improvement in the vertical ride with the modified damper material was not as much as expected. Further investigations were requested in order to establish the level of bedding in of the friction dampers and to compare the damping levels between the original and modified suspension types. These investigations included a series of wedge tests undertaken at the EDU workshop at the Railway Technical Centre. A wedge test was a method of determining the critical damping level of the suspension and was achieved by rolling all four wheels of one of the wagon bogies up shallow inclined wedges. As the wheels rolled over the end of the wedges they dropped simultaneously about one inch back down onto the rail creating a controlled suspension input. Instruments were fitted to record the acceleration levels at the axlebox and bogie frame and also potentiometers to record the suspension displacements. The data signals were conditioned via amplifier units and signals sent to a paper trace recorder to provide an instant output for analysis. The level of damping available (damping factor) within the suspension and the natural suspension frequency were determined by evaluating the rate at which the oscillation of the suspension decays after dropping of the wedge. A damping factor in the order of 0.25 vertically and 0.4 laterally are required to maintain stability.

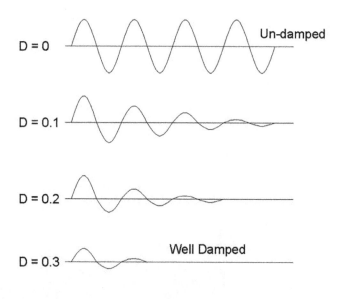

Fig.88 - Level of damping

The results of the wedge tests conducted on the freightliner container wagons fitted with standard friction damping material and the new manganese friction damping material confirmed the levels of damping were approaching the design values, and supported the theory that the new damping material was not fully bedded in at the time of the ride tests. A ride retest of the freightliner vehicle with the new damping material was subsequently recommended after the vehicle had gained some operational use in service.

Container traffic was on the increase through the 1980s, and to support this, a number of railway wagon leasing companies procured various types of new flat wagons; one of which was an 80 tonne GLW wagon type capable of carrying standard height containers of various lengths, 20ft, 30ft or 40ft on a 60ft load bed underframe. Running on GPS-20 type bogies, the wagons were built for Tiphook Rail by Rautaruukki in Finland, and designated TOPS code PFA with a designed operating speed of 75 mile/h. The first wagon of the new wagon type, TIPH93243 arrived at the RTC Derby in January 1987 for a full suite of static and dynamic acceptance tests. Dynamic test running commenced on the 14th January after completion of the static tests, with tare and laden ride tests between Derby and Cricklewood testing up to 85 mile/h, followed by slip/brake tests between Crewe & Winsford being completed before the end of January. The results of the ride and brake tests were analysed and I wrote the test reports for submission to the DM&EE Freight Design Engineers team. Whilst the riding and braking performance was deemed to be just acceptable, it was considered close to the limits for a new vehicle; therefore some minor modifications were made to the suspension springs, dampers and brakes system before the wagon was returned to the RTC for a second successful series of tests in August 1987.

Fig.89 - Tiphook 82 tonne container wagon, GPS-20 bogies, August 1987

The repeat tare ride test was conducted on the 10th August running from Derby to Finedon Rd (just north of Wellingborough), and returning via Lawley St (Birmingham) where the wagon was loaded with test weight containers at the Freightliner Terminal. Following a laden slip/brake test on the 13th August a repeat laden ride test was carried out on the 17th August again to Cricklewood and back. A final outing with the wagon was made on the 10th September on a short test run between Derby and Leicester. This series of three ride tests unusually saw a different locomotive type provided as traction for each test; Class 31, Class 37 and Class 47.

In contrast to the 60' long bogie container flat wagons, Procor built a batch of 40 tonne GLW, 2-axle PFA type wagons for use by Kellys with specially designed high capacity 20' long containers to carry coal. The unusual looking wagon PR93209, referred to as the 'roller-skate' by some of the team, arrived at the RTC in July 1987 and was quickly shunted onto the weighbridge for wheel weighing and torsional stiffness ($\Delta Q/Q$) testing. The wagon itself was relatively light, being only 11 tonnes, therefore allowing approximately 29 tonnes of load to be carried whilst maintaining a maximum 20 tonne axleload; however a wagon with such light tare weight provided challenges for suspension designers, in being able to negotiate track twists whilst maintaining sufficient levels of stiffness and damping to control the dynamic ride.

Fig.90 - PFA wagon PR93209 at the RTC, August 1987

This particular wagon needed some adjustment to the suspension following the initial torsional stiffness ($\Delta Q/Q$) tests and returned on the 12th August 1987 for a repeat test. Kellys special coal containers were required for the laden tests, because the underframe of the wagon was not flat, and the framework above each axle protruded about 6 inches above the load bed, which meant the underside of the containers had to incorporate an associated cut-out. Later in their operational life the wagons were transferred away from the coal container duties and modified with raised container mounting points to allow standard 20' containers to be carried.

The designs of freight vehicle suspnsions had not dramatically changed through the 1970s and 1980s; with a predominance in high friction content 25 tonne axleload designs that were not especially kind to the track on which they were operating. During the mid 1980s the British Rail Research (BRR) division developed a new, more track friendly, light weight frame low unsprung mass bogie design; incorporating in-board axleboxes, disc brakes, primary coil springs, rubber secondary springs and hydraulic damping. The bogie type was designated Low Track Force (LTF) and was designed to reduce the forces generated between wheel and rail during running compared to more traditional bogie types.

Fig.91 - BAA with LTF bogies at Cricklewood, May 1987

In early 1985 British Rail Engineering Limited (BREL) in Derby built a new BAA type 102 tonne GLW steel carrier wagon, (lot 4059); which was designated TOPS code YXA and numbered RDC921000. The wagon was used as the test bed for the new design of LTF bogie over a five year period between 1985 and 1990. Initial static tests were conducted by the DM&EE on the test rigs at the EDU workshop, however the first dynamic running trials were undertaken by the BRR department in March and April 1985 at the Old Dalby test facility, then continuing over many routes comparing the LTF bogie fitted wagon with a standard BAA wagon in varying load conditions up to 25 tonnes axleload at speeds up to 75 mile/h.

The further series of dynamic on-track tests was requested in May 1987, including DM&EE operated ride tests, in both tare and fully laden conditions, on the Midland Main Line route to Cricklewood with Test Car 1 and the ZXA Van. The laden condition was achieved using steel test weight blocks weighing half a tonne each, retained on the wagon using a wall of old railway sleepers along each side of the load bed. Although this method of loading was time consuming, it could be easily carried out within the EDU workshop without the need to take vehicles off-site for loading; it also enabled us to obtain very accurately the correct gross laden weight for the vehicle, by using the EDU weighbridge to validate the weight after loading and adjust as necessary.

Fig.92 - LTF *bogie tests with yaw dampers*

The YXA returned again in 1990 for further ride testing following modifications which included the fitting of yaw dampers and experimental RD12 wheel profiles. On the 4th May 1989, coupled with Test Car 1, a test run achieved 100 mile/h on the WCML between Crewe and Bletchley being hauled by a Class 86 locomotive. The final series of tests with the DM&EE were conducted between 22nd and 25th January 1990 and saw the LTF bogie fitted YXA wagon used for some axle strain comparison tests with a similar BAA wagon that was fitted with FBT6 bogies. Wagons RDC921000 (LTF) and BAA900283 (FBT6), both in the fully laden condition were instrumented at the RTC Derby with strain gauges on one axle of each wagon, and telemetry equipment to transmit the output signals from the rotating axles to the signal conditioning equipment in Test Car 4. Three days of test running were completed over a variety of routes including:-

- Peterborough to Reading via Cambridge and Stratford,
- Reading to Southampton, Salisbury to Reading via Westbury,
- Reading to Derby via Didcot and Birmingham,
- Derby to Barrow via Crewe and Carnforth,
- and Carnforth to Derby via Leeds.

After completion of the testing the instrumentation and strain gauges were removed from both wagons and the axles re-treated as necessary with the correct protection paint to prevent corrosion occurring where the strain gauges had been fitted. In between the periods of testing the YXA vehicle was used in steel coil carrying service trains with other BAA type wagons. The success of the LTF bogie design was clear by the large number of wagons produced during the 1990s that were fitted with a further development of this design of bogie, albeit operating at speeds up to 75 mile/h maximum.

Fig.93 - O&K 102 tonne box wagons test at Corby, March 1989 [Mike Fraser]

Foster Yeoman, having made significant investment in the UK with the Class 59 locomotives, progressed onto procuring a new fleet of aggregate wagons, commencing in 1988 with 102 tonne GLW box wagons. The wagons ordered were built by Orenstein & Koppel (O&K), a German engineering company specialising in heavy mechanical handling equipment (mainly for the quarrying industry), escalators and railway vehicles (mainly freight wagons and bogies). The 102tonne GLW box wagons designated TOPS code PTA were of a new design (code PH012A) fitted with DB Type 25-100, 2 metre wheelbase bogies mounted at 7.75 metre bogie centres and Brunninghaus 4 + 1 dual rate leaf springs. The designed maximum operating speed was 60 mile/h in both the tare and loaded conditions. This bogie type was a completely new design, based on the German standard DB662 variant, but with an extended wheelbase from 1.8 metre to 2.0 metre, and a higher axle load of 25.5 tonnes, especially for use in the UK. Two wagons were presented to the RTC at Derby for testing in November 1988; being outer vehicles OK3268 & OK3269 which were both fitted with standard buffers and drawgear at one end, and an AAR type knuckle coupler at the other end. When in service these two wagons formed the ends of a rake of wagons fitted with AAR couplers within the rake. The initial static tests carried out on both OK3268 and OK3269 showed the torsional stiffness properties to be well within limits but the hysteresis band indicating the friction levels in the suspension was fairly narrow; and the bogie rotational resistance 'X' factor results were slightly lower than expected.

When it came to the first dynamic on-track tests, the minor concerns raised during the static tests were unfortunately confirmed with a less than satisfactory ride performance; the vertical ride accelerations were outside the Freight Acceptance Curve limits on both welded and jointed track at 60 mile/h. The initial test was conducted on the 14th November 1988 over the Midland Main Line route between Derby and Bedford via Manton and Corby, and data was also collected during some special test runs undertaken as part of a bridge resonance study over Western Region routes on the 19th and 20th November. The lateral ride accelerations were also marginally acceptable and lateral instability was noted at speeds above 45 mile/h when running on both welded and jointed track types. This was not good news for Foster Yeoman or O&K, however the test team ploughed on with the assessment of the braking performance of the new wagons. There was some good news here as the brake stopping distance slip/brake test results were acceptable for operation up to 60 mile/h.

The results of the ride tests were analysed and the wagon designers considered that improvements could be made to the ride performance by the installation of UIC type sidebearers to increase the bogie rotation torque, increasing the 'X' factor test results and therefore hopefully reduce the tendency for lateral instability. In addition spring augment clamps were fitted to try and assist in increasing the friction damping available to control the vertical ride performance. These modifications were completed by March 1989 and wagons OK3271 & OK3272 were presented to the RTC Derby for further assessment. A set of repeat static tests were undertaken on 13th March 1989 and the wagons instrumented for further on-track tests which were conducted over the same Midland Main Line route between Derby and Bedford via Manton and Corby. The results of the repeat box wagon tests were promising in terms of the lateral ride accelerations, these being much better and the instability seen during the initial tests was not prevalent during the retests; however the vertical acceleration performance in the laden condition on the jointed track sections was still marginally failing the Freight Acceptance Criteria.

In June 1989 the second variant of wagons built by O&K for Foster Yeoman were ready for testing, these being a 102 tonne hopper wagon running of the same

DB Type 25-100, 2 metre wheelbase bogies. Ride testing was carried out between Derby and Bedford with the first run on the 19th June coupled to Test Car 1 with the ZXA van and hauled by 31416, returning to Derby via the Bardon Hill Quarry for loading with aggregate.

Fig.94 - Class 20s with O&K hoppers at Corby, June 1989 [Mike Fraser]

On the 29th June the test train ran again to Bedford this time hauled by two Class 20 locomotives, pausing at Corby en-route awaiting access to the line to Kettering. Exhibiting much the same ride characteristics as the box wagons tested in March, the O&K hopper wagons had issues at speeds above 45 mile/h.

Again the design engineers went back to the drawing board and later in the year in October 1989 box wagons OK3271 and OK3272 were presented again for testing, modified with the installation of Brunninghaus 4 + 2 springs, increasing the spring rate in the laden condition.

Vehicle Type	Vehicle Number	Instruments/Load
Locomotive	56055	-
PGA Wagon	PR 14336	Tare
PGA Wagon	PR14094	Tare
PGA Wagon	PR14002	Tare
Brake Van	B954636	Test Equipment base
PGA Wagon	PR14129	Tare
O&K Hopper Wagon	OK19300	Instrumented / Laden
O&K Box Wagon	OK3272	Instrumented / Laden
PGA Wagon	PR14176	Tare
PGA Wagon	PR14085	Tare

Table 11 - O&K Wagon test train formation, June 1990

The results this time were improved and the operating speed increased slightly. However Foster Yeoman had still not got the full 60 mile/h operational wagons they had desired. During June 1990 a further series of tests were conducted this time using a temporary test set up in an old (un-braked) brake van, based at Westbury yard. The train included an O&K box and a hopper wagon marshaled between a number of PGA hoper wagons which provided additional brake force.

Fig.95 - Yeoman wagons test at Westbury, June 1990 [Mike Fraser]

A number of combinations of different spring rates, with and without fixed and adjustable friction augment and spring link arrangements were trialed over a three day period between 22th June and 24th June 1990. The final outcome resulted in no significant improvement for either the box or hopper wagons and so they remained restricted to 55 mile/h tare and 45 mile/h when laden.

The operation in the UK of wagons originating from, and registered in Europe was common practice, however it was important to ensure that the running characteristics of such wagons met with the respective criteria for acceptance before they could be allowed to operate in the UK. In particular, attention was paid to lateral stability and the wheel profiles because in many parts of Europe the rail inclination and the rail head profiles are different. In the UK the rails are supported on sleepers secured to rail chairs that incline the rail-head inwards at an angle of 1-in-20, whereas in parts of Europe the rails are inclined at an angle of 1-in-40; therefore a different shapes of wheel profile are preferred in each case in order to maintain stability. A UIC registered 2-axle Tank Wagon number 23707490418-4 was just such an example that arrived for assessment of the riding properties in the UK in early 1990. This vehicle had 8 plate leaf springs, UIC double link suspension and was fitted with the P10 wheel profiles (the British Rail equivalent of the European S1002 wheel profile). A static torsional stiffness

(ΔQ/Q) test conducted on the 30th January showed the vehicle had a very stiff suspension and only marginal acceptability when negotiating track twists. The tare condition wagon was then coupled to Test Car 1 and instrumentation installed before heading out on a ride test on the 5th February that was originally planned to run to Bedford, however the poor vertical and lateral riding properties of the wagon meant that the test run was cut short and the locomotive ran-round at Kettering. After returning to Derby the vehicle was loaded using water using the hose-pipe outside the EDU Weighbridge, and a laden condition test run on the 8th February conducted between Derby and Kettering. Although the vertical riding properties when loaded were much better than when tare, the poor lateral ride meant that the test run was again cut short. The problems seen during the tests lead to operations of the wagons being speed restricted when in the UK. Further investigations were subsequently carried out on another wagon of the same design that was fitted with the British Rail wheel profile designated P5 which was preferred for use on freight stock in the UK. Two axle Tank Wagon number 23707390635-4 was provided for re-testing in October 1991, and test runs in both the tare and laden conditions were undertaken with Test Car 1 between Derby and Bedford. Although the lateral riding properties were improved as a result of fitting the P5 wheel profiles, the vertical ride was still did not meet the criteria therefore the wagon type had to remain speed restricted when operating in the UK.

Fig.96 - BAA steel carrier wagon 900172 coupled with Test Car 6, March 1998

In 1997 as part of the development of new designs of wagons for English Welsh & Scottish Railway (EWS), two 25 tonne axleload steel carrying wagons were modified with NACO 3-piece type bogies for trials. Wagons BBA910567 and BAA900172 were presented to the RTC Derby in October 1997 for testing and underwent the standard package of static torsional stiffness ($\Delta Q/Q$) and bogie rotation tests, then ride tests and finally slip/brake tests.

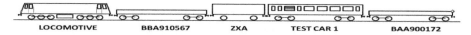

LOCOMOTIVE   BBA910567   ZXA   TEST CAR 1   BAA900172

Fig.97 - Steel Carriers with NACO bogies, test train formation

The ride test train was coupled in a different to normal configuration for these tests because we had two wagons to tests, and it was important that both wagons were tested at the trailing end of the formation at some point during the test. The BBA wagon was therefore coupled between the ZXA and the locomotive for the outward run when the BAA wagon was trailing; the BBA then being trailing for the return run. Dynamic test running began on the 6th November 1997 in the laden condition with Test Car 1 and a return test run to Wigan. The riding properties of the wagons were not as expected and the following day a series of wedge tests were undertaken at the EDU to take a look at the damping levels present in the 3-piece bogie friction dampers. Due to the design of the bogie it was necessary to conduct the wedge tests with four one-inch wedges that had to be perfectly aligned such that all four wheels dropped simultaneously off the wedges to be able to generate the required suspension input to determine the damping factor. The vehicles were then unloaded and a tare condition dynamic ride test to Warrington was carried out on the 12th November. Again after the tests further wedge tests were carried out in the tare condition, and following assessment of the results by the design engineers, a higher rated suspension spring and friction damping arrangement was installed into both wagons. The ride tests runs in both tare and laden conditions were repeated on the 14th and 20th November respectively in both cases coupled with Test Car 1 and operating a return trip to Warrington up to the designed operating speeds of 75 mile/h at tare and 60 mile/h when loaded to 102 tonne GLW. This series of tests was the last for which we used Test Car 1 before freight ride testing operations moved over to use Test Car 6 complete with the new on-board analysis (OBA) installation. At the time we were all so wrapped up with technology progress and Test Car 6 provided the best base for this; therefore the demise of the use of Test Car 1 went somewhat without ceremony which was quite sad really considering the old Great Western auto trailer coach had supported the dynamic ride testing of hundreds of wagons based out of the RTC Derby for nearly 30 years.

Fig.98 - BAA 900172, NACO bogies fitted with test instrumentation

The testing of the BBA and BAA wagons fitted with the NACO bogies continued into 1998 following various modifications to the suspension to improve the riding characteristics. Both vehicles were re-instrumented and coupled to Test Car 6 in early March followed by ride test runs between Derby and Carnforth on the 3rd March in the Tare condition and also between Derby and Warrington on the 9th March in the laden condition. Further investigations with the BAA wagon ballasted with an additional 5 tonnes on top of the normal tare weight were carried out between Derby and Carnforth on the 16th March and a final tare test run on the 20th March to Warrington with the BBA wagon only. Another series of modifications to the springs and the friction damper system on the NACO bogies were implemented and the BAA wagon 900172 was re-presented for ride testing again in May 1998. A tare test was run with Test Car 6 on the 21st May to Carnforth following which the wagon was loaded to 102 tonne GLW, and a laden condition test to Warrington carried out on the 28th May. The wagon was then released into normal service for a period of evaluation, but returned yet again in August 1998 for a tare ride test check run to Carnforth after completing 5000 miles in revenue traffic.

The first of the new Thrall Europa built BYA steel coil carrier wagons 966001 was delivered from York to the RTC Derby for test in June 1998, static tests were undertaken including bogie rotational resistance and also side push tests. The purpose of the side push tests was to determine the lateral stiffness hysteresis in the suspension that provided an indication as to whether the final series of modifications from the BAA wagon development trials had realised the desired effect on the production bogies fitted under the BYA coil carrier. After instrumentation fitting and coupling to Test Car 6, a laden condition ride test was carried out on the 2nd July 1997 across the normal route between Derby and Crewe, continuing up the West Coast Main line to Warrington. The locomotive normally ran-round in the down side goods loop adjacent to the station at Warrington which attracted some attention from the enthusiasts on the platform; especially on such occasions as this test, when a Class 37 locomotive (37245) was used to haul the test train. Upon return to Derby the parking brake test was carried out on BYA966001 followed by unloading to the tare condition. Between the 6th and 8th July the torsional stiffness ($\Delta Q/Q$) tests were undertaken on the weighbridge at the EDU followed by bogie rotation test on both bogies and a side push test in the tare condition. After re-coupling to Test Car 6 and connection on the instrument cables, checking the calibrations etc, a tare condition ride test to Carnforth was carried out on the 9th July. The wagon was the released into traffic to operate at 60 mile/h; the production batch of BYA wagons followed, entering service shortly afterwards.

Fig.99 - BYA steel carrier wagon, June 1998

BRAKING SYSTEM TESTS

STATIC BRAKE TESTS

Before embarking on any dynamic running tests the functionality and set-up of the braking system of new vehicles were checked statically to confirm the operation was in accordance with design and set-up criteria.

Fig.100 - WIA car carrier wagon, brake trolley static brake tests, May 1993

An example of these static tests is described below based on the two derivatives of the auto air train braking system fitted to most UK freight stock in the 1980s; these are often referred to as 'single pipe' or 'two pipe' braking systems.

The difference between a 'single pipe' and 'two pipe' brake system is the method of supplying air, and the pressure of the air from the hauling locomotive to charge the air brake auxiliary reservoir on each vehicle. In the 'single pipe' system the air to charge the reservoirs is fed via the control (brake) pipe at a maximum pressure of 5 bar, however when the driver operates the brake valve in the cab, this varies the pressure in the auto air brake pipe, thus providing a control signal to the distributors on each vehicle to apply air from the auxiliary reservoir into the brake cylinders. In the 'two pipe' system the air reservoir on each vehicle is fed via a separate main reservoir pipe which is maintained at a constant 7 bar by the locomotive; the brake pipe being used to provide the pneumatic control function to apply and release the brakes.

Single Pipe Auto Air Brake Wagon

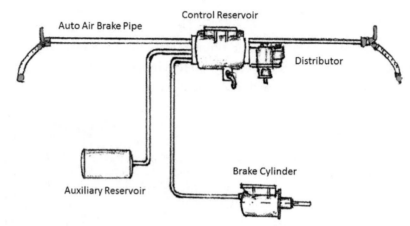

Fig.101 - Single pipe auto air brake system diagram

Two Pipe Auto Air Brake Wagon

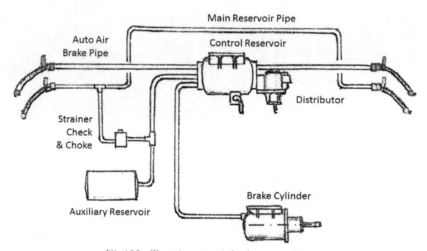

Fig.102 - Two pipe auto air brake system diagram

The diagram in Fig.103 shows the pressure measurements taken at the front and the rear of a 65 wagon train, firstly when connected in 'single pipe' mode, and secondly in the lower diagram showing the benefits of a 'two pipe' system; providing a much more even brake cylinder release throughout the train, particularly on longer trains. This allows for a quicker get-away and also helps prevent brake drag at the rear of a train when moving off. Where a vehicle under test was fitted with a 'two pipe' brake system, the static brake tests were carried out coupled in both 'single pipe' and 'two pipe' modes. The tests normally consisted of fitting calibrated pressure gauges into the brake cylinder(s) on the vehicle, and operating the brakes using a specially built brake test rig (often referred to as the brake trolley), to enable measurements of leakage rates and the brake force build up and release times to be recorded. During each brake application and release operation the rise and fall time of the brake cylinder pressure was timed using a stopwatch; this was normally repeated at least three times to ensure consistency of operation, and an average of the test results was compared with the limiting criteria.

There are two brake system timing control settings currently in use, denoted as Passenger and Goods modes; these settings for trains using either 'single pipe' or 'two pipe' mode, are designed to take account of the different inter-vehicle coupling arrangements, operating characteristics and operating speeds of freight and passenger trains. The difference between the Passenger and Goods settings is in the brake force application and release times, the Goods timings being slower in both cases. It is not normally permitted to operate with mixed Passenger and Goods settings within a single train as this causes mismatched brake operating characteristics between vehicles.

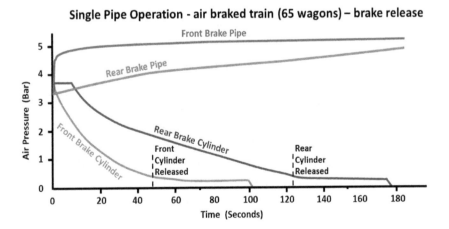

Single Pipe Operation - air braked train (65 wagons) – brake release

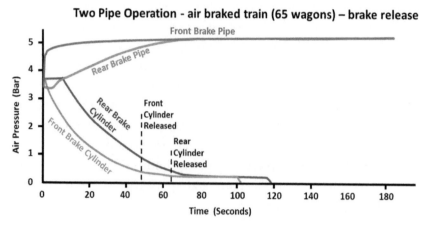

Fig.103 - Single/two pipe brake operation

The Goods setting is only used for vehicles in train formations operating up to 60 mile/h maximum, therefore freight vehicles with higher operating speeds are often fitted with a dual Passenger/Goods system.

Brake Timing Type	Application Time for Brake Cylinder Pressure to Reach 95% of Maximum (Seconds)		Release Time for Brake Cylinder Pressure to Fall from Maximum to 0.4 bar (Seconds)	
EMERGENCY	Single Pipe System	Two Pipe System	Single Pipe System	Two Pipe System
PASSENGER	3 – 5	2 – 3	15 – 20	15–20
GOODS	18 – 30	9–15	30–45	30–45

Table 12 - Static air brake cylinder application & release timings limits

The selection of the Passenger or Goods setting is made on each vehicle in a train manually using a changeover lever; the Passenger/Goods switch on the hauling locomotive was also set manually to match the train system setting. The criteria and methods applicable for undertaking static brake tests in the 1980s were contained in mandatory procedures defining the Requirements & Recommendations for the Design of Wagons Running on British Rail Lines, and the Specification of Brake Equipment for Air Braked Freight Stock. A 'two pipe' air brake system will revert to operate as a 'single pipe' system should the main reservoir pipe not be connected, and the auxiliary reservoir subsequently charged via the brake pipe to a maximum of 5 bar. An example of the differences between Passenger and Goods timings settings can be seen from the diagram in Fig.104.

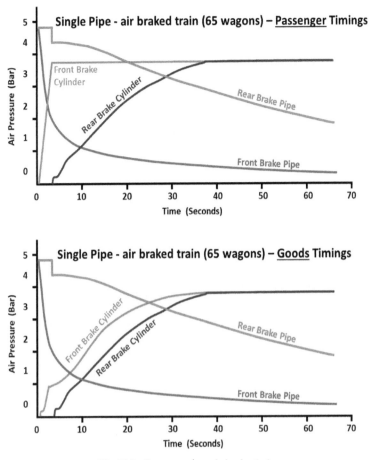

Fig.104 - Passenger / goods brake timings

The slower application of the air into the brake cylinders when in the Goods timing setting provides a very much more even application of the brake between the front and the rear of a long train, which helps prevent coupling snatching and shocks in long freight trains.

In addition to the statutory brake cylinder pressure application and release timings measurements, brake pad or block force measurements were often recorded in order to confirm design values were realised at the brake interface, and that brake effort was not lost in within deficiencies in the brake rigging or operating mechanisms between the brake cylinder and the brake pad or block. Specially designed load cells were used to undertake brake pad and brake block force measurements. After removing a brake pad or block, the load cells were positioned between the brake pad holder and disc, or the block holder and wheel, and the brakes applied; two readings were taken, one after the brake was first applied and a second reading after knocking the brake rigging (normally with a soft hammer or a brake stick) in order to simulate a moving vehicle condition; i.e. relieve any sticking in the brake rigging movement. Each test was conducted 3 or 4 times to ensure repeatability before removing the load cell, refitting the brake pad or block and moving on to the next wheel. In most cases only a sample of brake pad or block positions were recorded for each vehicle under test; however where any unexplained differences were noted between the results at each wheel position, then it was necessary to record all positions to obtain a full picture of the brake system operation.

The following table provides an example of a typical freight wagon static test measurements of brake application and release timings, in this case a 79 tonne GLW ballast spoil handling wagon which was fitted with a 'two pipe' air brake system operating one brake cylinder on each bogie and tandem cast iron brake blocks.

Spoil Wagon - Laden		Bogie 1	Bogie 2	Limits
Single Pipe (Passenger)	Brake Cyl. Pressure	3.72 bar	3.65 bar	
	Release Time	14.5 sec	13.2 sec	15–20 sec
	Application Time	4.2 sec	3.8 sec	3 – 5 sec
Single Pipe (Goods)	Brake Cyl. Pressure	3.79 bar	3.81 bar	
	Release Time	35.3 sec	34.6 sec	30 – 45 sec
	Application Time	27.0 sec	24.0 sec	18 – 30 sec
Two Pipe (Passenger)	Brake Cyl. Pressure	3.72 bar	3.79 bar	
	Release Time	14.0 sec	12.5 sec	15 – 20 sec
	Application Time	2.3 sec	2.4 sec	2 – 3 sec
Two Pipe (Goods)	Brake Cyl. Pressure	3.79 bar	3.88 bar	
	Release Time	35.2 sec	34.6sec	30– 45 sec
	Application Time	15.7 sec	14.8 sec	9–15 sec

Table 13 - Static air brake cylinder application & release timings test results

PARKING BRAKE TESTS

The performance of a vehicle parking brake (or handbrake) was determined by conducting a static test whereby the force needed to move a vehicle along the rails, with its handbrake applied, was measured and used to determine the gradient on which the parking brake would hold the vehicle. In order to move the test vehicle along the rails, a hydraulic ram fitted with a calibrated load cell (with 100 kN capacity) was installed between the drawgear of the vehicle under test and an anchor vehicle on the same line. The anchor vehicle was parked with its air brakes applied and wooden chocks placed tightly in front of at least two wheels. Although the actual load condition of the vehicle under test was not important for a parking brake test, it was always preferable to undertake a parking brake test with the vehicle in the loaded state to assist with adhesion between the wheels and rails, therefore ensuring rotation of the wheels during the test. A test was invalid if the wheels slid along the rails, therefore it was normal practice to carry out the test within a workshop such as the EDU (Engineering Development Unit) at the RTC Derby, where the rails were, in mostly cases clean, dry and level.

Fig.105 - An example of a lever type parking brake

To ensure a consistent and representative application of the parking brake on older design freight vehicles fitted with a lever type parking brakes, weights were attached the lever that were equivalent to the standard level of effort that could normally be applied at the lever end by the average person weighing about 60 kg applying the parking brake. On vehicle types where the parking brake was applied by a handwheel, a specified level of torque was used when applying the parking brake. This was calculated and applied with a calibrated torque wrench based on the diameter of the handwheel and the standard level of effort of 500 Newtons applied at the handwheel rim when applying the parking brake. On newer designs of rolling stock, electrically, hydraulically or spring operated/air released parking brakes were more common than using handwheels.

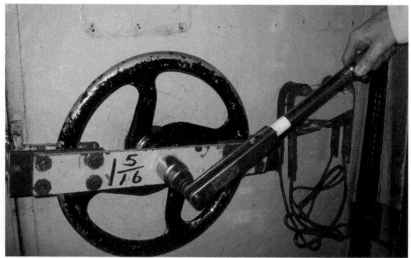

Fig.106 - Torque wrench used to apply parking brake handwheel on a locomotive

On freight vehicles it was common practice to have handbrake wheels fitted to both sides of the vehicle; however the operating mechanism to affect the application of the parking brake was often different on each side of the vehicle. This was because additional gears or levers were necessary to transfer the operation from one side of the vehicle. In such cases the test was always repeated to assess the performance of the parking brake when applied from both sides of the vehicle.

Fig.107 - Typical parking brake hand wheel on a freight wagon

The acceptability of the performance of a parking brake was assessed by establishing the gradient on which the vehicle under test could be held without rolling away. For example the criteria in place in the 1980s stipulated that a locomotive parking brake must be able to hold the locomotive stationary on a gradient of 1-in-30.

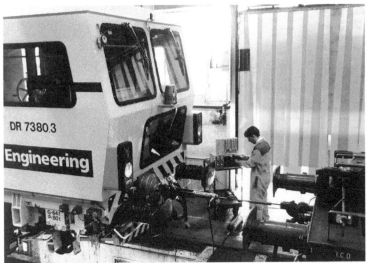

Fig.108 - Plasser tamper DR73803 Parking brake test at the EDU workshop [Serco]

For freight wagons and on-track plant vehicles, the performance of the parking brake was required to be capable of holding a fully laden vehicle stationary on a gradient of 1-in-40. In order to undertake each parking brake test, the hydraulic ram was operated either by a hand pump or by an electrically powered hydraulic pump in order to haul the vehicle under test for a distance of between 8 and 10 inches along the rails. During this movement the test engineer watched closely to ensure the wheels of the vehicle under test were rotating and not sliding along the rails and that the anchor vehicle remained secure.

Fig.109 - Example of a parking brake test on a Class 47 locomotive

The measurements taken from the load cell were amplified using the standard DM&EE purpose built signal conditioning units in order to provide a suitable analogue voltage output that was normally recorded onto a paper trace chart recorder. Post-test analysis would be confirmed back at the office. The peak value of resisting force was read from the chart manually and recorded onto the results table. After each test was completed it was necessary to return the vehicle back to its original position ready to start the next test. In the case of lighter weight vehicles with a low rolling resistance, this could be done by a couple of people pushing the vehicle, however for heavier vehicles or those with a higher rolling resistance, an atlas-bar (sometimes called a pinch-bar) was needed to rotate the wheels and get the vehicle moving. The equivalent holding gradient results for each test was then calculated as follows :-

$$Equivalent\ Gradient = \frac{Vehicle\ Weight(kN)}{Parking\ Brake\ Resistance(kN)}$$

Where:-

$$Parking\ Brake\ Resistance(kN) = Measured\ Resistance(kN) - Rolling\ Resistance(kN)$$

Test Vehicle	Test Wagon 'A'			
Test Date	- 1987			
Vehicle Weight (during test)	Laden (101.3 tonne)			
Vehicle Gross Laden Weight (Design)	102 tonne = (1000.3 kN)			
Parking Brake Type	Handwheel & Screw on both sides of the vehicle, directly applying on Side A, and indirectly via a reverse gear from Side B.			
Test	**Description**	**Measured Resistance (kN)**	**Parking Brake Resistance (kN)**	**Holding Gradient (1-in)**
Test 1	Rolling Resistance	4	-	-
Test 2	Handwheel Side A	34.1	30.1	33.2
Test 3	Handwheel Side A – repeat	34.8	30.8	32.5
Test 4	Handwheel Side B	29.7	25.7	38.9
Test 5	Handwheel Side B – repeat	30.1	26.1	38.3
Test 6	Rolling Resistance Check	4	-	-

Table 14 - Parking brake test results example

In the case of the example results for Test Wagon 'A' shown in Table 14, the parking brake performance was acceptable, with a calculated holding gradient of 1-in-38.9 worst case, which is better that the limiting 1-in-40 gradient. The parking brake applied from Side A of the wagon was seen to be better than that at Side B, this was expected because on this type of wagon the Side B handwheel applies the parking brake indirectly via a reverse gear.

The testing team were often called upon to help with incident investigations; one such occasion was in August 1994 when we undertook static brake tests including a parking brake test on Class 37 locomotive number 37113. The locomotive was severely damaged after running away from Edinburgh Waverley station on the 13th August 1994, and colliding with a stationary HST headed by power car number 43180. Myself and David Chorley collected the DM&EE test team's static brake test equipment, including a brake trolley, a hydraulic ram and calibrated load cell unit; loaded it all into the back of a Ford Mondeo estate car and set off to Scotland. The tests were undertaken in a small workshop at the Portobello yard near to the Craigentinny depot to the east of Edinburgh. We installed the equipment between the drawhooks of the undamaged No 2 end of 37113 and a second Class 37 locomotive number 37351 that was used as an anchor in order to undertake a parking brake test. The parking brake test was carried out satisfactorily using the handwheel in the No 2 cab, however the results of the test using the handwheel in the No 1 (damaged end) cab were adversely affected by the damage to the handbrake mechanism between the underframe and the bogies. Static brake tests with the air brake system were very also very limited due to the extent of the damage to the locomotive; however by installing pipe coupling adaptors in various positions in the air system we were able to apply an air feed to ascertain whether parts such as the air reservoirs and brake cylinders had leaks.

Fig.110 - 37113 awaiting brake system investigations, August 1994

The results from tests such as this, which provided evidence for accident investigations, were formally reported for the purposes of the investigation only. The badly damaged locomotive 37113 was subsequently withdrawn and cut-up in the yard at Portobello in Edinburgh.

Fig.111 - Parking brake test under way on 37113 (left), August 1994

BRAKE STOPPING PERFORMANCE TESTS

Dynamic brake testing of freight vehicles was predominantly carried out using the slip/brake method whereby the brake stopping distance performance of the vehicle was measured with the vehicle independent and not coupled with other vehicles in a train. In order to conduct slip/brake test a special coupling arrangement was used, the design of which originated from use on North Eastern Railway (NER) J21 and J25 type steam locomotives providing banking assistance on the Darlington to Kirkby Stephen line over Stainmore Summit in the 1940s and 50s. A wire operating cable was routed from the coupling mounted on the front drawhook, along the top of the boiler, and into the steam locomotive cab allowing the traincrew to uncouple the banking engines from the assisted train without stopping. Three of these coupling head units were obtained by the DM&EE in the early 1970s, adapted for use on slip/brake testing, and semi-permanently fitted to the heavy duty draw hook installed at the operating end (slip-end) of Test Car 2 (ADB975397). A spare slip coupling head was always carried on Test Car 2 along with the screw coupling, spare air brake pipes and brake pipe sealing rubbers.

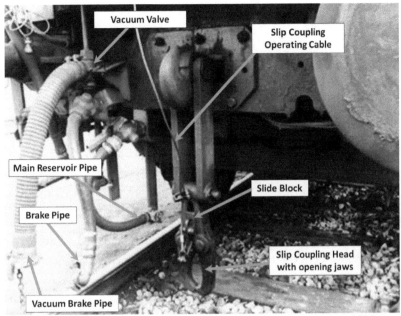

Fig.112 - Test Car 2 slip coupling and slip-end brake pipes arrangement, 1985.

The modifications to the braking system on Test Car 2 to facilitate use for slip/brake testing were initially made during 1974, such that during slip/brake testing the vacuum brake could be used for the locomotive and test coach braking, and an independent auto air brake system was provided for the vehicle under test supplied by the main reservoir pipe fed from the locomotive. This configuration also provided the ability to undertake slip/brake tests on vacuum brake vehicles by using the special self-sealing vacuum application valve, however by the early 1980s the requirement for such tests had fallen away due to demise in the of the use of the vacuum brake system on British Rail. The use of the special brake configuration on Test Car 2 was only permitted within the boundaries of an agreed slip/brake test site possession. After arriving at the test site the brake systems were configured for test and static test checks for correct operation were carried out. These static checks included measurement of brake cylinder application timings and pressures on the vehicle under test, and a continuity brake test on the locomotive and test car brakes. Once the static brake tests were completed to the satisfaction of the Officer in Charge (OIC), the slip coupling was fitted and testing could commence. After testing was completed the brake system was configured back in the normal manner and the statutory brake continuity test completed before the train was hauled out of the test site.

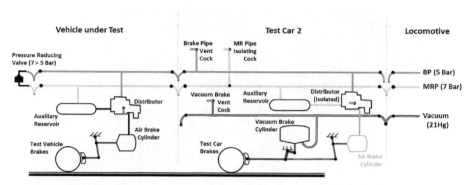

Fig.113 - Brake schematic for a 2 Pipe slip/brake test, dual brake locomotive

With reference to the diagram in Fig.113, representing the brake system set-up used for a 'two pipe' vehicle slip/brake test, with a dual brake locomotive; Test Car 2s brakes were operated in the normal manner by the Vacuum pipe from the locomotive. The Main Reservoir Pipe was also connected between the locomotive and the test car, but the Brake Pipe was not connected. The Vacuum Brake Vent Cock was installed at the slip-end of Test Car 2, providing the OIC of the test with the ability to vent the Vacuum Pipe in to apply the locomotive and test car brakes if required. The test vehicle brakes were operated by the OIC of the test from within Test Car 2, independently of the loco/test car brakes; the air supply for the test vehicle brakes being provided by the Main Reservoir Pipe at the standard 7 bar from the locomotive. The pressure reducing valve mounted at the rear of the vehicle under test reduced the 7 bar Main Reservoir Pipe air to 5 bar, and fed into the Brake Pipe and release the test vehicle brakes. During a test, after the slip coupling was operated, the Officer in Charge (OIC) of the test closed the Main Reservoir Pipe to prevent draining the locomotive air supply, then vented the brake pipe which applies an emergency application to the test vehicles brakes.

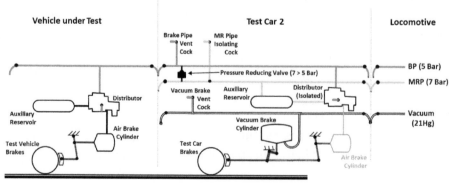

Fig.114 - Brake schematic for a 1 pipe slip/brake, dual brake locomotive

When testing a single pipe brake system fitted test vehicle as represented in Fig.114, the pressure reducing valve fitted at the slip-end of Test Car 2 was used, being connected between the Brake and Main Reservoir Pipes, provided the correct 5 bar brake pipe pressure. With the demise of the use of vacuum brakes on British Rail, and the fall in the number of dual braked locomotives, Test Car 2 was modified during the early 1990s to allow its air brakes to be operated (whilst in slip/brake test mode) in the normal manner by the Brake Pipe from an air brake only fitted locomotive, the Test Car Brake Pipe Vent Cock providing the Officer in Charge (OIC) of the test in Test Car 2 with the ability to apply the locomotive and test car brakes if required.

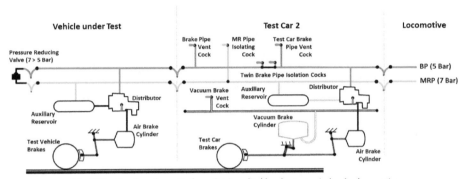

Fig.115 - Brake schematic for a 2 pipe slip/brake test, air brake locomotive

The modifications as represented in the diagram in Fig.115, included twin Brake Pipe isolation cocks providing a safety dead section in the test car brake pipe, allowing the test vehicle brakes to be operated by the Officer in Charge (OIC) of the test from within Test Car 2 independently of the locomotive and test car brakes; the air supply for the test vehicle brakes being provided by the Main Reservoir Pipe fed from the locomotive.

The slip/brake tests were predominantly carried out on the Down Slow line between Crewe and Winsford north of Crewe Coal yard, within T3 possession (single line blocked to all traffic, protected by stop boards and detonators at each end of the possession section). The test section of line was particularly suited to the slip/brake testing because the test section was level (no gradients), and the four track formation allowed normal traffic to continue whilst we were testing. It was not permitted to carry out slip/brake testing during hours of darkness, thick fog or falling snow and at least 1 mile clear visibility was needed before testing could commence; this was because the Officer in Charge (OIC) had to be able to clearly keep the test vehicle in sight at all times when the vehicle was detached from the test car. On a few occasions I remember a whole day of testing being cancelled upon arrival at Crewe because of thick fog. A dual channel independent intercom system with battery backup was installed in Test Car 2 to provide dependable means of communication between the Officer in Charge (OIC) and the locomotive driving cab. Before the start of each test, heavy duty instrument cables fitted with Plessey 6 way connectors were attached between the Test Car, along the outside of the locomotive and into the leading cab (normally through the drivers cab-side window), and connected to twin intercom speaker/microphone outstations. The locomotive was always manned by a driver and a traction inspector for slip/brake testing; this allowed the Officer in Charge (OIC) to communicate directly with the traction inspector in the cab so as not to distract the driver from his duties. The test train formation was normally a locomotive + Test Car 2 + the test vehicle, with the locomotive at the north end of the train; and in most cases the trains operated out and back from the Derby RTC for each day of testing. The purpose built Time-Speed-Distance-Unit (TSDU) in Test Car 2, that was used to record slip/brake test brake stopping distance data, was supplied with a TTL pulse signal from an axle-end mounted toothed wheel pulse generator. The settings of the TSDU were always checked before the start of slip/brake testing during the transit movement from Derby, whereby the distance measuring counter was started when passing a milepost and stopped when passing the next milepost. In general the mileposts on the lineside were reasonably accurately located; however to reduce the level of inaccuracy the distance checks were normally checked over at least 5 miles. An accuracy of +/- 10 feet in 5 miles (<0.1%) was achievable and considered sufficiently accurate for undertaking slip/brake testing. The TSDUs and the Pulse Generators received regular bench calibration checks to confirm the speed function accuracy, however the TSDU set-up, the relationship between speed and distance, and the pulse generators were very reliable and did not need resetting often.

We normally departed Derby RTC at around 8am, pausing on route at Crewe station for a traincrew change and entered into the T3 possession at Coal Yard just after 10am. The possession arrangements and information regarding the slip/brake testing operations were published in the Weekly Operating Notices (WONs) for the Crewe to Winsford route section such that drivers of service trains and more importantly any track maintenance or patrolling staff were aware of the testing. We commenced testing after setting up the intercom system between the test car and the locomotive, fitting the distributor release peg protection plate, conducting static brake pressure and timings checks on the test vehicle, fitting the flashing safety indicator lights to the test vehicle and installing the slip coupling. The first test run was always a Down Slow line proving run and to check the north end of the T3 possession, that the stop board and detonators were in place and also to make sure that no track maintenance staff were on or about the line.

Fig.116 - A view from the drop-down slip-end window during testing [Mike Fraser]

Operation of the slip/brake test was effected by accelerating the train up to the pre-arranged test speed, and at the designated slip point at Coppenhall (passing under bridge number 7, just north of 161 milepost) the slip coupling was operated using the wire cable to open the coupling jaws; once the coupling had been released then the brakes of the vehicle under test were applied by the Officer in Charge (OIC). As the vehicle brakes started to apply the gap opened up between the test coach and the vehicle; the brake pipe stretched slightly and pulled apart, the locomotive and test coach continued onward, the movement indication lights helping the Officer in Charge (OIC) determine when the vehicle had stopped. The instruction was then given to the traction inspector via the intercom to the cab that the vehicle under test had stopped and the loco/test coach could be brought to a stand. At all times the Officer in Charge (OIC) was keeping a watchful eye on the vehicle under test, ensuring that it remained stationary; once the test stopping distance information had been recorded, the OIC then instructed the traction inspector to reverse the locomotive and test coach back along the Down Slow line to collect the test vehicle. After the vehicle under test was re-connected to the test coach with the slip/coupling and a brake operation check carried out, the whole formation was propelled back towards Crewe up the Down Slow line for a sufficient distance in readiness to commence the next test run.

Fig.117 - Slip/brake flashing safety indicator lights unit on test wagon [Mike Fraser]

The following series of pictures takes us through the process of a slip/brake test at bridge number 7, on the Down Slow line north of Crewe, in this case a test of 'single pipe' air braked fitted PHA 90 tonne hopper wagon RHR17301, hauled by a dual braked Class 47 locomotive number 47188 on 6th July 1987.

Fig.118 - Slip/brake 1 - approaching the test site

The whole train formation was accelerated to the pre-determined test speed in the normal running direction along the Down Slow line north of Coal Yard. Just before the slip-point the driver shut the locomotive power handle and the train was allowed to coast, the OIC was advised by the traction inspector via the intercom system that the test could commence and provided a verbal countdown to advise the OIC when the locomotive was passing under the bridge. The OIC ensured all persons in the test car were ready; the OIC was ready at the slip-end next to the brake pipe vent cock, the coupling operator standing at the slip-end drop down window with the operating cable in-hand, and the equipment operator at their seat in the instrument area (and not in the kitchen making the tea) to ensure the purpose built Time-Speed-Distance-Unit (TSDU) distance counter was set to zero.

Fig.119 - Slip/brake 2 - power off

Shutting the power off on the locomotive allowed the buffer gap between test car and the vehicle under test to close up slightly, making it much easier for the coupling operator to release the slip coupling.

Fig.120 - Slip/brake 3 - start the test, apply the brakes

When passing under bridge number 7 (just before 161 ¼ milepost) which was the designated slip-point, the coupling operator released the slip coupling, and once satisfied that the slip coupling had dropped clear of the test vehicle drawhook, the OIC applied the brakes on the test vehicle using the brake pipe vent cock. The TSDU distance counter automatically started after receiving an electrical signal from a reed switch mounted on the brake pipe vent cock. It was however possible for the OIC to partially move the brake pipe vent cock handle which still applied the test vehicle brakes, but did not trigger the reed switch; therefore the equipment operator always watched the TSDU to ensure the trigger to start the counter from the brake pipe vent cock had operated correctly. The OIC observed to ensure that the brake pipes had pulled apart cleanly and then close the main reservoir pipe isolating cock to prevent excessive drain on the locomotive air reservoir supply.

Fig.121 - Slip/brake 4 - vehicle starts to slow down

As the brakes applied fully on the test vehicle it started to decelerate, the locomotive and test car continuing forward, coasting along the down slow line. It was important at this point that the driver did not apply the locomotive brakes as this could cause the vehicle under test to run into the back of the test car. As the slip/brake testing was always carried out during the daytime, and the adjacent main lines were open to normal traffic, often the drivers of down direction trains passing whilst a slip/brake test was in progress would, if they had not seen before, look in amazement as the test vehicle parted from the test coach, giving the OIC a nervous wave and a smile as they passed.

On one occasion I remember a driver of a passing train actually stopping his service train further up the line and make an emergency phone call to the signaller to report that a train on the down slow line had divided.

The OIC in the test coach kept constant watch of the vehicle under test ensuring that it continued to decelerate and regularly informing the traincrew via the intercom system of the status whilst the test vehicle was still moving.

Fig.122 - Slip/brake 5 - under observation

To assit the OIC in determining whether the test vehicle was still moving, the safety indicator lights unit fitted to the end of the vehicle under test changed from flashing white lights changing to a steady red light, only when the vehcile had stopped.

Fig.123 - Slip/brake 6 - vehicle slowing to a stop

This particular test was carried out using a dual braked locomotive, therefore the locomotive brake selector setting was in Vacuum Passenger such that the vacuum train brakes were operational on the test car. The intercom system twin cables running to the locomotive leading cab can be seen in Fig.123 attached to the special underframe cable brackets on the test car and along the side of the locomotive, carefully routed under the cab step to prevent a trip hazard for the traincrew.

Fig.124 - Slip/brake 7 - keeping a watchfull eye

Once the OIC had assured himself that the vehicle under test had stopped, then the traincrew were instructed to bring the locomotive and test car to a stand. The distance measurement was taken and the counter re-started by the equipment operator. The OIC then instructed the traincrew to propel Test Car 2 back up the Down Slow line to collect the test vehicle, at all times the OIC keeping a watchful eye on the test vehicle. The train guard was normally with the OIC at the slip-end of the test car during the propelling move, and after the test car had buffered up to the test vehicle the distance counter was stopped and the test result calculated. The slip coupling was re-attached by the guard helped by the coupling operator and a brake operation check conducted before the whole formation was reversed up the Down Slow at 20 mile/h maximum to the start point ready to repeat the test process again. The test equipment operator calculated the test vehicle brake stopping distance for each slip/brake test run, taking the total distance from the point at which the brakes were applied to the vehicle under test until the locomotive and test coach had stopped, minus the distance the locomotive and test coach had to reverse back to collect the vehicle under test. A deceleration meter fitted on the headstock of the vehicle under test containing a calibrated accelerometer and a 0.1 Hz low pass filtered output onto a strip paper chart, provided an accurate measure and characteristic of retardation rate of the vehicle under test throughout each slip/brake test.

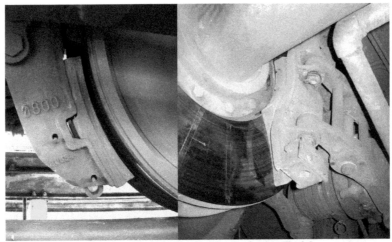

Fig.125 - Tread brakes [Left] - disc brakes [Right]

The design of the braking system, the weight of the vehicle and the type of brake pad or block used could result in very different deceleration characteristics during a brake stop. Cast Iron brake blocks were used in the majority of applications until mid-1960s, however the braking characteristic of the cast iron block is such that fade is prevalent as the speed increases.

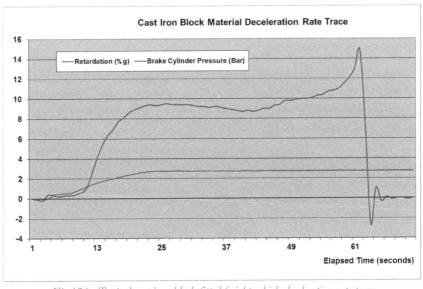

Fig.126 - Typical cast iron block fitted freight vehicle deceleration rate trace

As the requirements grew to run trains faster on a regular basis, the need to stop them more effectively and consistently, promoted the development of the disc brake for rail vehicle applications. The disc brake with the use of composition pad materials proved more effective at higher speeds with less fade and a more consistent braking rate throughout the speed range. As can be seen from the typical cast iron brake block fitted freight vehicle deceleration rate trace in Fig.126, the brake effectiveness faded during the brake stop, but the retardation rate increased at slower speeds where sliding was more likely to occur; whereas a composition brake pad fitted disc braked vehicle produced a more consistent rate of deceleration throughout the brake stop as shown in Fig.127.

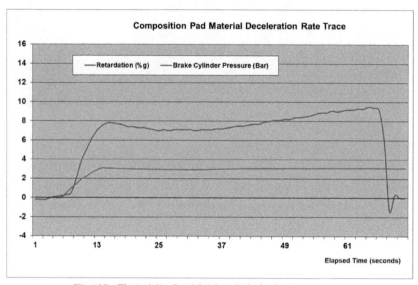

Fig.127 - Typical disc fitted freight vehicle deceleration rate trace

Slip/brake tests were always carried out at varying speeds from 30 mile/h up to the maximum operating speed of the vehicle under test. In order to minimise the effects of unrepresentative temperature increase due to repeated brake stops, the test speeds were varied, normally alternating between lower and higher speeds.

PARAMETER	DESCRIPTION
Type of Brake Application -	The brake setting Passenger or Goods and whether the vehicle had a 'single pipe' or 'two pipe' brake system fitted.
Brake Demand -	Always Emergency for slip/brake testing.
Initial Speed -	The speed at the slip-point when the brake was applied to the test vehicle.
Forward distance -	The distance measured from the slip-point when the brake was applied to the test vehicle to the point at which the test car stops.
Reverse distance -	The distance from the location where the test car stopped, running back to collect the test vehicle.
Distance -	The calculated brake stopping distance for the vehicle under test.
Mean Retardation -	The average deceleration rate of the vehicle brake test that was calculated.

Table 15 - Parameters logged during slip/brake testing

On average between 10 and 18 slip/brake test runs could be completed during a day's testing, however the achievable number of tests varied depending on how slick the team on board was working, the type of locomotive, the performance of the vehicle under test, and more often than not, the weather. During the slip/brake testing, the brake stopping distance measurement results were recorded manually on template log forms.

DYNAMIC SLIP/BRAKE TEST RESULTS

Project Title : Test Wagon 'A' - Brake Performance						Location : Crewe - Winsford			
Vehicle Condition : Laden (102 tonne Gross)						Rail Condition : Dry			
Train Formation : Class 47 + Test Car 2 + Test Wagon 'A'						Test Date : - 1987			
Test No.	Brake System & Setting	Brake Demand	Initial Speed (mile/h)	Test Car Forward Dist (ft)	Test Car Reverse Dist (ft)	Test Vehicle Dist (ft)	Test Vehicle (metres)	Mean Retardation (%g)
1	1 pipe Pass	Emergency	29.7	2047.4	1611.8	435.6	132.7	6.74
2	1 pipe Pass	Emergency	51.3	3918.9	2538.9	1380.0	420.4	6.35
3	1 pipe Pass	Emergency	76.8	6708.7	3261.1	3447.6	1050.4	5.70
4	1 pipe Pass	Emergency	46.9	3473.5	2391.1	1082.4	329.8	6.77
5	1 pipe Pass	Emergency	66	5408.5	3031.3	2377.2	724.3	6.10
6	1 pipe Pass	Emergency	75	6451.6	3248.8	3202.8	975.8	5.85
7	1 pipe Pass	Emergency	37.4	2651.3	1982.9	668.4	203.6	6.97
8	1 Pipe Goods	Emergency	59.5	5055.8	2617.4	2438.4	742.9	4.83
9	1 pipe Pass	Emergency	43.1	3138.1	2232.1	906.0	276.0	6.83
10	1 pipe Pass	Emergency	59.2	4674.4	2832.4	1842.0	561.2	6.34
11	1 pipe Pass	Emergency	53.4	4088.0	2637.2	1450.8	442.0	6.55
12	1 pipe Pass	Emergency	60.2	5130.0	2638.8	2491.2	759.0	4.84
13	1 Pipe Goods	Emergency	67.4	5567.1	3067.5	2499.6	761.5	6.05

Table 16 - Example slip/brake test results

The test results were plotted on a graph to show how they compare to the criteria, which in this case was the British Railways Standard, Braking System and Performance for Freight Vehicles that was applicable in the mid-1990s.

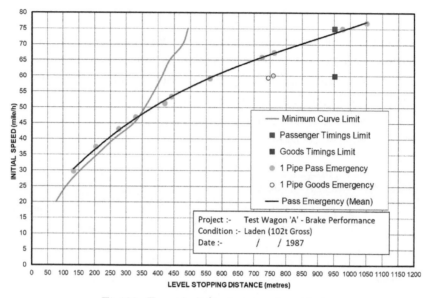

Fig.128 - Example slip/brake test results graph plot

It can be seen from the example test results in Fig.128 that this particular wagon failed to meet the criteria on two counts, firstly the brake stopping distance performance at 75 mile/h exceeded the 951 metre limit; and secondly the brake stopping distances at speeds up to 47 mile/h were less than the minimum recommended curve. Although the minimum limit curve was not mandatory, it was a recommended minimum performance level in order to limit wheelslide and therefore reduce the likelihood of wheel tread damage. When unacceptable results were obtained from a slip/brake test it normally resulted in an investigation and often modification to the vehicle braking system, followed by a re-test.

The number of projects on the go at any one time for the freight test team was varied; sometimes as many as four vehicles of different types were at the RTC on test or being prepared for test. In late January 1986 this was just such a case, with the Iron Ore Tippler bogie stress project wagon (PTA26100) in preparation, there was also a Plasser LWRT Power Wagon (DR89010) part way through its acceptance test programme, a Railease Container Flat (RLS92648) for slip/brake tests and also an Autic (MAT) 3 axle car articulated carrier wagon (RLS92051), again for slip/brake Tests.

The Autic wagon underwent static brake testing during January, including measurements of static brake cylinder pressure application and release timings, brake block forces and a handbrake test. Unlike ride tests where there was a considerable amount of instrument cables connections between the Test Car and the vehicle under test; slip/brake testing did not have any cable connections between the vehicle and the test coach therefore the vehicle was often attached to Test Car 2 for slip/brake testing on the morning of each test run. A days slip/brake testing was normally operated each test day running out and back from the Railway Technical Centre in Derby. The formation had to be in the correct orientation before leaving Derby so that Test Car slip-end was facing south when the train was running on the Crewe to Winsford test possession site.

Fig.129 - Autic (MAT) car carrier RLS 92051 at Winsford, February 1986

The locomotive was attached to the test car and the Class 08 shunt locomotive on the north end of the train on the Way & Works siding. The Class 08 hauled the formation into Derby Station where it detached allowing the test train locomotive to take us on to Crewe via North Stafford Junction (on the Birmingham line south of Derby) and Stoke-on-Trent. Arriving at Crewe the normal practice was to pause for a traincrew change in the station before heading north towards Coal Yard on the Down Slow line, where the traction inspector walked across to the signal box to confirm the T3 possession arrangements with the signaller, and collect the trackman with his detonators and stop boards to set-up the T3 possession. The test train then moved forward into the possession site and the detonators and stop boards protection placed behind the train. Once in place we proceeded right the way to the north end of the possession to check the protection was in place, and also check for any staff working along the line in the vicinity of the test location. The test car and vehicle under test were then propelled back along the Down Slow line to the south end of the possession ready to start the brake days testing. For the Autic Car Carrier wagon the first days brake testing was on a cold 27th January 1986, with the wagon in the tare (no payload) condition, then again on the 13th February in the loaded condition, with steel weights secured on the car carrier decks with timber framing to simulate the cars. Slip/brake testing in the winter months was always a cold affair because of the need to have the slip-end drop-down window on Test Car 2 open to operate the slip coupling; however there were plenty of heaters fitted inside the slip-end and instrument area to help. Once we had completed the days slip/brake testing the train was re-coupled and a full train brake continuity test done by the traincrew before releasing the T3 possession and heading north off the Down Slow line. In order to get the train back to Derby it was then necessary to run the locomotive round to the south end; this was normally carried out near Winsford in the old Over & Wharton Salt Works reception siding loop, however in later days after the closure of these sidings the train ran to Hartford or Warrington to allow the locomotive run-round before heading south back to Crewe and onward via Stoke-on Trent to Derby. On arrival at Derby station the RTC Class 08 shunt engine coupled onto the south end of the test train and drag the formation back into the RTC yard where the locomotive was detached and released to return to No 4 shed at Derby or sometimes Toton or even back to Crewe Diesel depot.

Another standard slip/brake test was conducted on a Sheerness scrap steel wagon in the tare (no payload) condition in April 1986. The wagon, number PR3150, was built by Procor using reclaimed and refurbished bogies from life expired 102 tonne tank wagons, and a new high-sided box body. Although the static brake tests (including brake cylinder pressures) undertaken at Derby were considered in line with new design expectations; the slip/brake tests did not go well, with the stopping distance performance just exceeding the specifications. The wagon was returned to Procor and underwent some modifications to the braking system; returning for a repeat tare condition slip/brake test on the 10th July 1986. The locomotive provided for this day of testing was a newly painted Class 47 number 47190 in large logo grey livery. The brake stopping distance performance after modification was much improved and within the specified criteria. Due to availability of test slots it was some time later in the year before the Sheerness scrap steel wagon returned for a loaded condition (102 tonne GLW) slip/brake test; and on 2nd October 1986 locomotive 31191 hauled Test Car 2 and the wagon from Derby to Crewe where a locomotive exchange was effected, attaching Class 85 number 85007 to provide traction to undertake the slip/brake testing. The use of an electric locomotive helped tremendously for slip/brake testing because the acceleration rate was higher, therefore the distance and subsequently the time needed to get to speed for each test was much less than with a diesel locomotive, especially with a loaded wagon; which resulted in more test runs being achieved within a day of testing.

Fig.130 - Sheerness scrap steel PXA wagon PR3150 at Winsford, July 1986

The majority of the slip/brake testing was conducted on the West Coast Main Line between Crewe and Winsford, however on a few occasions the Down Main lines of the ECML between York and Thirsk was used when higher testing speeds were required. In June 1985 as part of the design preparations for the proposed IC225 trains, two Mk III coaches was fitted with different designs of axle mounted disc brakes were presented for test to the RTC Derby, the first vehicle being a Mk III HST TGS vehicle W44029 with standard BT10 bogies and fitted with Lucas Girling design brake discs and Ferodo brake pads. The second vehicle being a Mk III HST TO vehicle W42257 with BT35 bogies and fitted with Knorr Bremse design brake discs and Becorit brake pads. In order to allow coupling with Test Car 2 for the slip/brake testing standard buffers and drawgear was installed on both vehicles in place of the normal tightlock centre couplers used in HST formations.

Fig.131 - Slip/brake test on the ECML, June 1985

Testing commenced in each case with a series of static tests to record the brake system characteristics for each coach along with the brake pad force readings. The dynamic testing of each coach was undertaken individually to assess the performance of the different designs of brake system. The first coach W44029 being coupled with Test Car 2 at Derby on Saturday 22nd June 1985 and moved to the carriage sidings just to the north of York station where the train was stabled overnight.

Fig.132 - 47221 and Test Car 2, June 1985

At dawn on the Sunday the main lines north of York between Skelton Junction and Thirsk were closed to other trains to allow the normal method of slip/brake testing to be undertaken. The train was moved into the test section and a full series of slip/brake tests up to 95 mile/h were undertaken running north on the down main line; the slip point being at Tollerton. The maximum testing speed was constrained by the Class 47 locomotive.

Fig.133 - W42257 after a slip/brake test, standing on the ECML, June 1985

The slip/brake test flashing safety indicator lights provided the Officer in Charge (OIC) of the test with a visible indication that the coach had stopped after being slipped.

Fig.134 - A view from within the Mk III coach during a slip/brake test, June 1985

On these tests we also had brake cylinder pressure and deceleration rate recording equipment installed in the Mk III coach. Test engineers (including myself) took turns between operating the slip coupling in Test Car 2, in riding in the Mk III coach during the slip/brake tests to operate the recording equipment. By late morning testing had been completed and the main lines were opened to normal traffic and the test train was transit moved back to Derby. On the 29th June the same process was undertaken with the second coach W42257 and testing again completed on north of York on Sunday 30th June 1985. The results of the tests were reported back the design engineers working on the new brake system designs for the new IC225 project. The photos taken on the 30th June provide a unique insight into slip/brake testing, showing Test Car 2 in action and the rather unusual sight of a Mk III coach standing on its own in the middle of the East Coast Main Line.

Fig.135 - 47221 and Test Car 2 re-coupling to the Mk III coach, June 1985

Freight vehicles often arrived for acceptance testing prior to the application of a final livery. In March 1986 one of the first of a batch of new 90 tonne aggregate hopper wagons built by W H Davis for Bardon Hill quarries, commenced standard slip/brake testing whilst still in a green undercoat. Following the initial tests some minor brake system modifications were necessary to realise an acceptable brake stopping distance performance.

Fig.136 - PHA Bardon Hill wagon 17102 at Winsford, March 1986

Whilst the modifications were being carried out, the vehicle was painted in its smart Bardon Hill quarries livery before undergoing a final slip/brake tests.

Fig.137 - PHA Bardon Hill wagon 17102 at Ditton, May 1986

The primary reason for slip/brake testing was to establish the braking performance of the vehicle under test without influence from other vehicles. However in some cases vehicles could not be operated on their own which meant that they had to be tested whilst coupled with other vehicles. An example of this situation was the NFS wagons that operated as part of the High Output Ballast Cleaner (HOBC) train, and featured specially designed conveyor units that fed ballast from the hopper wagons into the Swivel Conveyor Wagons. When the NFS wagons were uncoupled from the Swivel Conveyor Wagons, the ballast conveyors fitted one end of each NFS wagon overhung the end of the vehicle, therefore special barrier wagons (converted from 2-axle open wagons) were needed to enable the NFS wagons to be moved separately. During the slip/brake testing between January and April 1995, the NFS wagon under test number DR92223 was coupled to its special barrier wagon during the tests. The brake stopping distance performance measured was only valid for the coupled formation of the NFS and the barrier wagon; therefore the barrier wagon itself had to be tested on its own, in order that an evaluation of the brake stopping distance performance of the NFS wagon itself could be derived from the results of both sets of tests.

Fig.138 - NFS wagon during slip/brake testing, April 1995

An example of results from tests undertaken at speeds up to 75 mile/h for a test wagon coupled with a barrier are shown in Fig.139, the graph was plotted including the results from the barrier vehicle only tests and the derived stopping distance performance for the test wagon on its own.

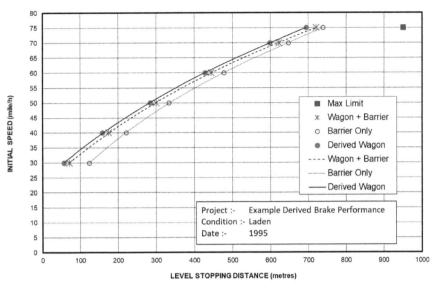

Fig.139 - Example brake test results actual & derived (without barrier wagon)

In this example the braking performance of the barrier vehicle was worse than that of the wagon under test, therefore the derived brake stopping distance performance for the test wagon was slightly better than that recorded during the test with the barrier vehicle coupled.

WSP SYSTEM TESTING

With the introduction of the disc brake systems and increased normal operating speeds, the overall deceleration rates of passenger trains under normal full service brake operation could also be increased. As a result of the enhanced deceleration rates, the probability of wheel slide during braking increased, potentially causing wheel tread damage and increased brake stopping distances. With the aim of combating this problem, wheel slide protection (WSP) systems were fitted to ensure that during high brake demand when adhesion levels were low, the wheelsets do not lock up. The WSP system compares the rotational speed of all the wheelsets on each vehicle, and using an independent reference speed controls the individual wheelset brake pressures to prevent lock-up whilst maintaining the brake effort demanded by the driver. In conjunction with brake stopping distance tests on newer passenger rolling stock the functionality of the WSP system needed to be proven. In some cases the brake system manufacturer requested a full suite of tests to establish the overall performance of the WSP system in accordance with the stringent European criteria in addition to the functionality and brake stopping distance performance under low adhesion conditions.

Fig.140 - WSP test train at Winsford, June 1988

An example of WSP testing that I was involved in was undertaken for Lucas Girling during mid-1988, when a new design of WSP control system was fitted to a Mk III locomotive hauled coach number 17174. The tests were conducted on the West Coast Main Line in the normal manner for slip/brake testing using Test Car 2. The Mk III coach was fitted out with test instrumentation and adhesion reducing equipment in the form of large water barrels and pipe-work to feed adhesion reducing solution in a controlled manner through nozzles positioned in front of both wheels of the leading wheelset. Additional pipe-work was installed to feed water in a controlled flow onto the brake discs and brake pads to simulate adverse wet weather conditions. Before commencing the dynamic testing, the brake cylinder pressures, air suspension pressures, brake cylinder pressure build-up, release times and brake pad loads were recorded at the RTC Derby. The Mk III coach was then coupled to Test Car 2, and to provide additional brake force during the transit movement from Derby to Crewe, RDB975428 and Test Car 1 were also included in the train formation. The first test running was carried out on the Down Slow line between Crewe and Winsford on the 6th June 1988, with the whole formation in order to bed-in the newly fitted brake pads, following which RDB975428 and Test Car 1 were uncoupled and stabled at the south end of the T3 possession near Crewe Coal Yard. Full service and emergency application brake tests were then carried out using the slip/brake method at a nominal test speed of 75 mile/h on dry rail. Following the dry rail tests, further full service and emergency brake tests were carried on a rail wetted only with water fed from the on-board barrels. The dry and wet brake performance was established on the first day of testing and the train formation was then re-coupled and returned to Derby via Winsford. The barrels were topped up and the train headed over to Crewe again the following day to continue with the low adhesion testing programme. A solution of environmentally friendly soap mixed with the water was prepared in the barrels such that the concentration of low adhesion fluid applied via the on-board system could be gradually increased with each brake stop until a level of adhesion was achieved between 6%g and 8%g. The adhesion level was determined from the rate of vehicle deceleration recorded at the point of initial wheel slide. The testing continued until the 24th June in order to fine tune the brake system settings to obtain the optimum brake stopping distance performance when braking in the controlled low adhesion conditions. The train was stabled on most nights between test days in Crewe to reduce the travelling transit movement time, although the train did return to Derby to the RTC between the 17th and 20th June. A considerable amount of test instrumentation was required to support a full functionality WSP system test, including measurements of the vehicle reference speed, all axle rotational speeds, all brake cylinder pressures, brake supply reservoir pressure, brake stopping distance and deceleration rate. In addition to the vehicle reference and axle speed information, we fitted an independent doppler radar in order to confirm the true vehicle speed during the test. An example of a typical output from a four axle vehicle WSP brake stop is shown below in Fig.141.

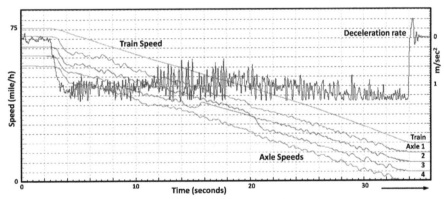

Fig.141 - Example WSP test time history trace

The rotational speeds of the axles can be seen to fluctuate as the WSP system reacts to the wheels starting to slide and adjusts the brake cylinder pressures to apply and release the brakes in order to maintain brake stopping distance performance without allowing the axles to lock-up and potentially cause wheel tread damage. Should a wheelset lock-up during a low adhesion brake test then the risk of damage to the wheel treads was high, therefore throughout WSP testing it was important to regularly inspect the wheel treads of the vehicle under tests for signs of damage or flats. Rectification of wheel flats normally meant a visit to a wheel lathe that could severely disrupt the course of a test schedule and booked test runs; luckily there were no such issues during the testing in June 1988. After the completion of the WSP testing and before we handed back the T3 possession, it was extremely important to ensure no traces of the low adhesion fluid remained on the rails of the Down Slow line that could adversely affect the brake stopping performance of service trains. A number of slow speed movements back and forth over the brake test area were therefore undertaken whilst spraying water onto the rail head. After re-coupling of the additional brake force vehicles, a confirmation full service brake test was then carried out during last train northbound run of the day, before heading out of the possession and onwards to Winsford.

RAIL VEHICLE STRUCTURAL ASSESSMENT

The design engineers at DM&EE often requested the testing department to assist in the verification of the structural integrity and fatigue life predictions of new designs, modifications and service failures. Such investigations entailed the test engineers working closely with the design and structural experts to develop the testing process dependent on the specific requirement of the verification. During January 1986 a series of tests were requested to investigate bogie frame fractures on British Steel PTA type 102 tonne Iron Ore Tippler wagons. The wagons were built by BREL at Shildon 1972 and had therefore already seen about 13 years in service. These tests involved the fitting of 24 strain gauges to the British Rail designed FBT3 fabricated bogie frames in order to evaluate the stress levels in various areas of the bogie frame. The series of tests included undertaking running tests starting at Derby heading north via Doncaster to Scunthorpe and Immingham that also incorporated the loading and unloading of the wagon whilst still recording the bogie frame stresses. This meant that Test Car 1 had to be positioned very close to the unloading tippler so the cables to the instruments could remain connected whilst the wagon was tipped upside down for unloading, this was a fantastic experience and not something many people have seen that close up (the large observation end windows in Test Car 1 making a great viewing platform for these tests).

Fig.142 - PTA iron ore tippler wagon 26100, February 1986

The installation of the strain gauges involved preparing the surface of the bogie frame, removing the paint down to the bare metal, and sticking the foil backed gauge to the metal surface using special super strong adhesive. The idea was that as the metal bends then the strain gauge also bends with it which causes the electrical resistance of the strain gauge to change. This change in resistance was then measured and equated to the level of strain (ε) in the metal. Strain gauges could only be used once, because it was not possible to remove them intact once fitted. The strain gauges were fitted to the bogies of wagon PTA26100 at the RTC Derby using tried and tested methods starting with the cleaning of the work area. The preparation of the surface area for each of the strain gauges started initially using a small grinder to remove all paint and defects such as hair line cracks, weld splatter and corrosion. For some of the less accessible strain gauge positions, it was not possible to access with the grinder so the tedious task of rubbing with abrasive papers was the only answer to get the area prepared. The size of the area to be cleaned for each strain gauge was about 50mm square, with a final surface finish obtained by rubbing with 400-grade silicon carbide paper. Once the suitable quality of surface finish had been achieved, it was rigorously cleaned with a mild acidic conditioner using cotton buds and tissues. A final single wipe of the cleaning medium was required because rubbing backwards and forwards risked further contamination of the surface, followed by a surface clean wipe using an acid neutralizer ready for attaching the strain gauges. Prior to attaching the strain gauges an identification of the exact gauge position was achieved using an old ball point pen to burnish the surface within the area surrounding the gauge position being careful not to mark the area where the proposed strain gauge was to fit.

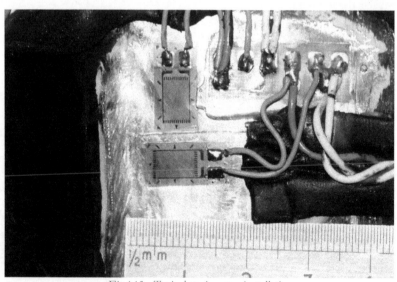

Fig.143 - Typical strain gauge installation

The strain gauges used for this particular test on the bogies were ¼ inch type with a nominal 120 Ω resistance, supplied sealed in plastic boxes in individual plastic wallets. Tweezers cleaned with proprietary cleaner were used to lift each gauge out of its wallet, and place individually on the clean surface of the inside face of the plastic strain gauge box lid, next to a solder tag strip, being positioned close to but not touching each other. A strip of Mylar tape was then pressed over the gauge and tags before they were transferred on the Mylar tape from the plastic box lid and positioned on the prepared test area being temporarily held in position with the Mylar tape in line with the pre-marked location. Preparation for sticking the gauge to the test surface was by peeling the Mylar tape back and applying a special strain gauge glue catalyst liquid to the underside of the gauge surface and solder tag using a small brush. The catalyst was left to dry for a couple of minutes before applying two drops of the special strain gauge adhesive and the Mylar tape then pushed down to secure the strain gauge and tags in position, holding with thumb pressure for a minute until the glue was set. The Mylar tape was then peeled away slowly to reveal the solder tags and the connecting wires attached using a small point tipped soldering iron.

The strain gauge area was then protected with a special non-conductive rubberised paint followed by an application of a thin mastic strip. In addition because these strain gauges were attached on bogie frames further mechanical and waterproof protection was applied using rubber pads inserted above and around the gauge area and wires, followed by aluminium adhesive tape and silicon sealant. As you can see the process for strain gauge installation involved cleanliness and precision which was no problem if you were working on a test bench, however this became significantly more challenging when fitting strain gauges to the least accessible corners on the underside of a bogie frame of a well-used iron ore carrying wagon. Connection of the strain gauge to the signal conditioning equipment was via the testing section standard shielded 6-core cables fitted with Plessey type connectors. An example of an electrical schematic of how each strain gauge was connected into a Wheatstone bridge arrangement is shown in Fig.144. In this case with one active gauge and the opposing legs of the circuit made up using resistors. It was normal practice for resistor 'C' to be positioned close to the strain gauge 'A' to create a half bridge configuration minimising interference in the cable connection back to the signal conditioning amplifier unit where the other half of the bridge, resistors 'B' and 'D', were provided by in-built resistors.

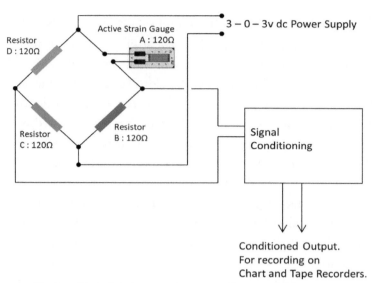

Fig.144 - Typical single active strain gauge Wheatstone Bridge diagram

The signal conditioning equipment used during these tests incorporated a number of fixed value calibration resistors to provide a shunt calibration facility whereby a known resistance value could be connected across one leg of the Wheatstone bridge circuit (usually the one containing the strain gauge), to simulate a known level of output from the measuring strain gauge. Based on the expected micro-strain ($\mu\epsilon$) levels on the bogie frames, suitable levels of signal for recording were selected for the dynamic tests with an initial setting of $100\,\mu\epsilon$/volt. It was necessary then to set the output voltage from the signal conditioning accordingly for recording onto the chart and data tape recorders. This was achieved by applying the shunt calibration and adjusting the amplifier gain until the required output voltage was reached.

On these tests with the Iron Ore Tippler wagon PTA26100, additional instrumentation was also installed, with accelerometers being fitted to the vehicle underframe and bogies to monitor the ride of the wagons and any track inputs during the tests; and potentiometers fitted across the suspension springs to monitor movements and the rotation angle of the bogies. The data from all of these transducers was monitored on-board the test coach but analysed post-test to evaluate the bogie stress levels and whether the track inputs, ride characteristics or loading operations were affecting the bogie frame stress levels.

Fig.145 - Interior of Test Car 1, January 1986

The typical 1980s scene inside Test Car 1 shown in Fig.145 with Dave Fearn scrutinising a UV paper chart record whilst preparing for the Iron Ore Tippler wagon tests. Note the small monitor used for displaying the camera image looking at the leading wheelset/rail interface and the digital speedometer readout showing 100 mile/h, this was an applied calibration check and not an actual speed as the test coach was stationary at the time of the photo. The data signals from the accelerometers, potentiometers and strain gauges were conditioned using purpose built amplifier units mounted in the instrument racks in the test car. Each data channel had a separate amplifier module that provided a power supply to each strain gauge and amplification of the returned data signals plus a capability to filter the data signals prior to recording. The output from the amplifier of each data channel was recorded onto magnetic tape using one inch reel-to-reel type magnetic tape recorders, in this case two types were used, a Honeywell 14-track recorder in the instrument rack and a more modern SE7000 type 14-track recorder secured onto the bench. Also on the bench area were three SE6300 Ultraviolet (UV) paper trace recorders used by engineers to assist in on-line monitoring the data recorded. The data signals from the amplifiers units were fed to the UV recorder internal galvanometers that reflected a directed light source onto the light sensitive paper producing a time history of each strain gauge output. The portable lamps on the bench were used to assist in exposure of the light sensitive UV recorder paper and reveal the traces; although the charts could not be left exposed to light for too long otherwise the whole chart was overexposed and the paper trace records lost forever.

Fig.146 - Iron ore tippler wagon test at Immingham, February 1986

The test runs commenced on the 1st February 1986 with Test Car 1, PTA26100 and an RFQ barrier wagon number B954935 hauled by a Class 47 between Derby and Scunthorpe. The barrier wagon was then uncoupled and a further three tippler wagons added to the test train. PTA26092, PTA26012 and PTA26096. On the 2nd February four test runs were completed as follows:-

Test Run 1 - Scunthorpe to Immingham, all four PTA wagons in the Tare condition,
Wagons 26100, 26092 and 26012 were loaded at Immingham to fully laden and the wheel weights checked with a portable wheel weight comparator device,

Test Run 2 - Immingham to Scunthorpe, three PTA wagons loaded, one remaining in Tare,

Test Run 3 - Scunthorpe to Immingham, three PTA wagons loaded, one remaining in Tare,

Test Run 4 - Immingham to Scunthorpe, three PTA wagons loaded, one remaining in Tare,
Wagons 26100, 26092 and 26012 were unloaded at the Scunthorpe BSC tippler depot.

Upon completion of the unloading at Scunthorpe the test instrumentation was removed from the wagons and PTA26100 uncoupled from Test Car 1. The wagons were released to return to normal service and Test Car 1 was hauled back to Derby. After the dynamic testing was completed the data tapes sealed in their dust protective cases, and chart roll records sealed in the dark in boxes, were returned to the analysis laboratory at the RTC Derby for post-test analysis. During the mid-1980s the DM&EE testing development team invested in a new SES (Stress Engineering Services) digital interface unit and along with new digital analysis facilities meant that it was possible to use the rainflow cycle counting technique (which was widely regarded as the industry standard) to make fatigue life predictions from recorded stress/strain time history data. The fatigue life calculations were produced by rainflow method cycle counting of the damaging stress cycles followed by an evaluation against the pre-defined weld classification for the particular area of the bogie structure, to produce and estimated life expectancy in terms of operational mileage.

Fig.147 - Foster Yeoman PHA 102 tonne hopper wagon GPS25 bogies

A warm August in 1986 saw the test team head down from Derby to Merehead Quarry to assist Foster Yeoman with further investigations on the 102 tonne aluminium bodied PHA type hopper wagons built by Procor. This time it was to investigate the emerging wagon body structural issues. A series of tests were undertaken after first fitting of strain gauges to various parts of the aluminium body structure of two wagons then dynamic testing in both the tare and loaded conditions over typical operating routes. The testing also included data collection whilst the wagons were being loaded with aggregate at Merehead Quarry to evaluate the stress levels in the body structure. In preparation for the tests, two PHA 102 tonne hopper wagons were delivered to the RTC Derby and in addition to the normal ride instruments installed on both wagons, a total of 55 strain gauges were applied and connected via instrument cables into Test Car 1. The locations of the strain gauges on the wagon body structure were prescribed by the DM&EE structural engineers' team. Preparation works took over two weeks to complete, and this particular test was the occasion I remember as the greatest amount of test equipment being installed within the saloon area in Test Car 1; for the purpose of these tests only, two additional 6ft high instrumentation racks installed on the left side of the saloon area to accommodate the signal conditioning amplifiers.

Fig.148 - Interior of Test Car 1 preparing for 102 tonne hopper stress tests, August 1986

Fig.148 shows Ray Smith and Dave Fearn in the very busy instrumentation saloon area and the test benches on the right side where two of the large one-inch format data tape recorders were installed, one 14-track and the other a 42-track machine, and five SE 6300 type UV paper chart recorders. Once the test instrumentation was installed and the signal conditioning amplifiers set up with the correct calibration settings for each data channel, a proving test movement was carried out within the RTC sidings hauled by the resident Class 08 shunt locomotive. All was in order and the following day, early on the 5th August 1986, a Class 31 locomotive number 31305 arrived from Derby 4-shed for a prompt 9am departure; the test run heading south along the Midland Main Line via Manton and Corby to Cricklewood recess sidings.

Fig.149 - En-route from Derby to Westbury, 31305 at Cricklewood, August 1986

After a short break and change of driver the ensemble continued on its way via Dudding Hill Junction and Acton Wells Junction onto the Great Western Main Line towards Reading, then via Newbury to Westbury Yard where the train was stabled overnight with Test Car 1 amidst rows of PGA hopper wagons.

Fig.150 - Test Car 1, at Westbury yard, August 1986

Having spent the night in a hotel in Westbury, on the morning of the 6th August we headed back to the Test Car in Westbury Yard and started the generator to warm up the test equipment ready for the day's testing. Locomotive 47131 was provided from Westbury shed and coupled to Test Car 1 for the short trip south to Witham East Somerset Junction and then along the single line branch to the Foster Yeoman terminal at Merehead. One of the main reasons for taking the test train to Merehead was so that the structural assessment measurements from the strain gauges fitted to the wagon body structure could be assessed whilst the wagons were being loaded. The Class 47 locomotive was therefore detached in the sidings at Merehead and the Foster Yeoman resident General Motors built locomotive No 44, Western Yeoman II was used to shunt the test train into the loading area at the Torr Works quarry. During the movements into the loading area and whilst both the wagons were loaded with aggregate, the recording equipment was running throughout. We had to remain inside the test car therefore pictures were not possible of the loading process.

Fig.151 - 47131 at Merehead with Test Car 1 and two PHA wagons, August 1986

Following the completion of the loading process the test train was drawn back into the Merehead sidings where the Class 47 was re-attached. Testing continued with a number of return trips running between Merehead and Westbury to gather data with the wagons running in the laden condition over the branch line. The day ended with the test train being stabled back in Westbury Yard and we retired to the hotel again in Westbury. The next day on the 7th August we had an early start to warm up the test instrumentation ready for the return trip back to Derby. The route via Newbury to Theale and Acton being representative of that the wagons saw in normal service when fully loaded, therefore providing good data for assessment by the structural engineers. Following return to Derby the instrumentation was removed and the wagons returned back to service. The data analysis could take anything from a few days to a few weeks depending on the workload within the analysis laboratory and the complexity of the analysis required. In this case the fatigue life assessment work from the strain gauge data was relatively straightforward and the test report was prepared within a couple of weeks for submission to the design and structural engineers for assessment.

An example of a more complex test installation is shown below in Fig.152 where strain gauges were fitted to the axles of an LTF25 type bogie of the new Amey Roadstone Corporation (ARC) 102 tonne aggregate hopper wagon SRW17901 in order to measure axle bending stresses. A series of running tests to measure the axle strains were carried out with the wagon in various load conditions in conjunction with the statutory ride tests over the Derby to Cricklewood, Midland Main Line route between the 10th and 18th October 1988. The wagon was loaded during the return leg of the test runs at the small ARC stone terminal to the north west side of Loughborough station. The new design of LTF type bogie being fitted with axle bearings in between the wheels meant that the lineside hot-axlebox detection equipment installed on many routes were not able to detect if a bearing was to run hot on this type of bogie. Therefore this prototype wagon was fitted with a new design of on-board hot axlebox detection system that used wax plugs installed into an air pipe on the bogie that melted the event of an axle bearing running hot, causing air to leak out of the pipe which in turn caused a valve in the brake system to operate applying the train brakes.

Fig.152 - LTF25 bogie strain gauge telemetry equipment to measure axle bending

Axle bearing temperature measurements were also taken during the tests on the Midland Main Line to provide assurance that the on-board hot axlebox detection system wax plugs were not operated during normal running. The axle bending stress test installation used telemetry technology adapted by the DM&EE testing development section for use on rail vehicle axles to transmit the strain gauge data signals from the rotating axles. This data was then recorded using the conventional tape data recorders in the test coach. Full bridge strain gauge configurations (with 4 active gauges) were installed on the axle at four locations along the length of the axle. Each full bridge data channel required power which was provided by a battery pack and a signal transmission module connected to a disc aerial. The battery packs and modules were bolted to base plates secured opposite to each other around the axle by stainless steel band straps. It was important to have the battery packs and modules installed opposite each other to maintain the weight balance whilst the axle was rotating. The disc aerials were manufactured specially from plastic discs in two halves that were bolted together around the axle. An aerial wire was then run from each signal transmission module round the respective disc to create an aerial. The transmitted data signals were then picked up using antennae positioned close to the disc aerials which were mounted on a steel bracket installed between the bogie side frames. The antenna were then connected back to the signal conditioning equipment in the test coach by using standard coax cables with BNC type connectors.

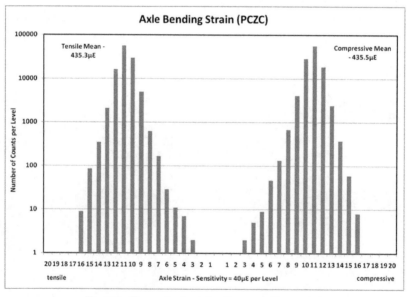

Fig.153 - Example of axle bending stress data analysis

Accelerometers and potentiometers were also installed on the bogie frame and axleboxes to measure suspension movement and track inputs that could be useful information when assessing the axle bending stress levels during the tests. An example of the output of an axle bending stress data analysis using the Peak Count Zero Crossing (PCZC) method is shown in Fig.153. In this example of a 3 hour 45 minute test section including traversing tight curves with check rails and also plain line running at speeds up to 60 mile/h, the even distribution of the strain levels and the low amplitudes were not cause for concern.

ON-TRACK PLANT VEHICLE TESTING

The maintenance of the tracks on the network was supported greatly in the 1970s and 1980s with the introduction of new types of on-track plant machines capable of doing the work of many men very much quicker. Such machines provided ballast movement, cleaning and regulating and also tamping which is a process whereby Tamping "Tines" powered by the machine packs the ballast under the sleeper to maintain a more stable sleeper bed and ensure the alignment of the track is maintained. The mechanical equipment on these machines associated with maintaining the tracks did not often feature in the testing undertaken by the DM&EE testing section, however all new or modified machines were presented for acceptance testing to ensure they complied with the mandatory vehicle running criteria. The design of the suspension and running gear of on-track plant was often very different to freight wagons or other rolling stock due to the nature of the operation once the machines were on-site and maintaining the track. For example some Tamping machines had hydraulic suspension blocking systems to prevent unwanted suspension movement during the tamping process. The static wheel weighing, torsional stiffness ($\Delta Q/Q$) and bogie rotation tests often involved obtaining assurance that the machines additional equipment did not have any adverse effect on the ability to negotiate track twists or curves. This chapter takes a look at a selection of different types of on-track plant vehicles that were presented to the DM&EE for testing.

BALLAST AND TAMPING MACHINES

In the mid-1980s Plasser & Theurer developed a new Pneumatic Ballast Injection (PBI) Machine designated PBI-84 and numbered DR73700. The PBI was initially tested in 1984 before I joined the test section however a further assessment of the ride comfort performance was required after a period in service. I took the trip down to the Plasser UK factory in West Ealing on the 7th October 1986 and undertook a test run with the PBI under its own power up to 45 mile/h with portable ride equipment between West Ealing, Reading and Newbury.

Fig.154 - Plasser PBI-84, DR73700 at West Ealing, October 1986

The test results highlighted a minor issue with the suspension and following some modifications DR73700 was subsequently tested again to confirm the ride characteristics for operation in-train formation. These tests were successfully carried out at speeds up to 60 mile/h coupled to Test Car 1 in two test runs on the Midland Main line between Derby and Kettering on the 28th January and between Derby and Finedon Road on the 29th January 1987. An example of a new design of tamping machine was presented in mid-1987 from Plasser & Theurer being an 09-16 CSM Tamper/Liner, TOPS code ZWA-M, numbered DR73001. This particular tamping machine was designed for continuous tamping of plain line track and had an additional single axle trailer at one end which contained track alignment equipment as well as the fuel tank, general equipment storage and load area. The torsional stiffness ($\Delta Q/Q$) test on this machine had to be undertaken in stages on the 23rd June 1987, because the overall length including the additional trailer was greater than the weighbridge at the RTC Derby. Following successful completion of the torsional stiffness and bogie rotational resistance tests, the normal test instrumentation set-up was installed including underframe and suspension mounted accelerometers and potentiometers to record the ride performance. Additional accelerometers were also fitted onto the cab floors and drivers seats to assess the vibration environment acceptance for traincrew.

A doppler radar device was installed on the underframe to provide a pulse signal for driving the independent speedometer and distance counter required to measure the brake stopping distance performance. The signal conditioning and data recording equipment was installed in the rear of the trailer-end cab in the tamping operators area. Portable sound pressure level meters were also used during the tests in both cabs to confirm the noise levels were within limits. In order to provide sufficient running time to capture data to demonstrate compliance for ride, braking and traincrew environment, two dynamic test runs were undertaken on the 29th and 30th June 1987, the first to Kettering and the second as far as Manton Junction. After entry into service the 09-16 CSM Tamper / Liner number 73001 worked on the eastern region for many years and after withdrawal from service and storage for a while, the machine was exported to work in South Africa in 2010.

Fig.155 - Plasser 09-16 CSM tamper at the RTC Derby, July 1987

Another new track maintenance machine developed by Plasser & Theurer was the Dynamic Track Stabiliser (DTS) which was designed to provide controlled, accelerated settling of the track following tamping, rather than relying on the passage of service trains to settle the track. The first of the new DTS 62-N machines, number DR72201 was submitted or testing in June 1987. Following completion of the statutory static testing, the DTS underwent a ride test on the 18th June coupled to Test Car 1, hauled by a Class 31 locomotive. Braking tests followed a few weeks later on 7th July hauled by Class 45, 45041 from Derby to Crewe with Test Car 2. After arrival at Crewe, the locomotive was uncoupled and self-powered testing of DR72201 undertaken between Crewe and Winsford on the Down Slow line. Once the testing had been completed, the DTS ran north to Winsford under its own power, returning along the Up Slow to Crewe where it was re-united with Test Car 2 and 45041 for haulage back to Derby.

Fig.156 - DTS 62-N at Crewe ready to start braking tests, July 1987

The cab area vibration environment assessment was retested following a period in service and on 29th February 1988 a further self-powered test run was carried out between Derby and Manton Junction using DTS 62-N machine DR72202.

A new Plasser & Theurer 08-16 SPI tamping machine number DR75001 was submitted for testing in mid-1991. I was involved in the static Delta Q/Q test on 29th July at the RTC and although I was on holiday when the ride test was carried out in August, I did manage to take a photo of the rather strange looking single cab machine in the RTC sidings awaiting the start of testing.

Fig.157 - Plasser 08-16 tamper DR75001, RTC Derby, July 1991

In June 1993 the Kershaw built High Output Ballast Cleaner (HOBC) classified TOPS code ZWA-D and numbered DR76101 was submitted for testing to the RTC Derby. Weighing in at about 275 tonnes, and over 130 metres in length the HOBC was an articulated six vehicle unit riding on seven VNH-1 type bogies. Three of the bogies had a hydraulic drive system fitted that was designed to allow self-powered operation at speeds up to 60 mile/h. The HOBC was also designed to be hauled in-train at speeds up to 75 mile/h being fitted with disc braking. Static wheel weight measurements and torsional stiffness ($\Delta Q/Q$) tests were undertaken in two stages due to the length of the HOBC being longer than the siding length to the rear of the EDU weighbridge, requiring the whole train being turned by a transit movement to Trent Junction between each set of tests. The bogie rotational resistance tests were concluded in one stage as the bogie rig was located towards the north end of the EDU workshop number 3-road, and with the traverser in place at the rear of the workshop there was sufficient length to accommodate the HOBC. Further static tests were conducted to demonstrate the compliance of the sound pressure levels of the warning horns fitted for use when the HOBC was operating in self-powered mode. The horn tests were carried out with the HOBC positioned on the RTC Way & Works sidings as there were no buildings or other structures nearby to reflect the sound of the horn. The clear area around the end of the train could be realised in accordance with the criteria, and the tests conducted on a Saturday morning so as not to disturb too many people working in the RTC offices. Instrument fitting for ride performance assessment was undertaken and coupled to Test Car 1, a dynamic test run was carried out on the 23rd June 1993 between Derby and Cricklewood hauled by 47972 at speeds up to 85 mile/h. After return to Derby the instrumentation was removed and the following day the HOBC without Test Car 1 was hauled again by 47972 on three return trips between Derby and Leicester for brake stopping distance performance assessment at 5 mile/h increments between 40 and 75 mile/h.

The HOBC entered service for a while then subsequently underwent extensive modifications resulting in the need to be re-presented for testing in mid-December 1996 when repeat static bogie rotation and torsional stiffness ($\Delta Q/Q$) tests were completed. After a further period in service DR76101 received more modifications, returning to the RTC in November 1997, then again in March/April and October 1998 for further tests that included bogie rotational resistance, warning horn sound pressure levels, braking and vibration dose environment tests in the cabs. The machine re-entered service again, operating for another 4 years following which it was stored and subsequently scrapped in about 2009.

Fig.158 - HOBC, DR76101 bogie rotation test, December 1996 [Serco]

RAIL GRINDING MACHINES

In the early 1980s Speno supplied Rail Grinding machines which were operated under contract on the UK network to provide rail head conditioning and re-profiling. Prior to their use on the network the normal statutory static and dynamic tests were carried out by the DM&EE at Derby. After joining the test section in 1985 I was involved with the static torsional stiffness ($\Delta Q/Q$) testing and on-track running testing of a 2-axle Switch and Crossing Grinder type URR 16P-46, number DX79500.

Later that same year, the first of the type URR48/4 Rail Grinding trains for plain line grinding operations arrived for testing. The four vehicle URR48/4 Rail Grinding train consisted of vehicles DR79211, 79212, 79213 and 79214. We undertook self-powered ride and braking performance tests over the Midland Main Line route up to speeds of 60 mile/h, with a British Rail driver overseeing and route conducting the Speno machine driver throughout.

Fig.159 - Speno URR48/4 rail grinding train at the RTC Derby, 1985

Five years later in February 1990, the next generation of Speno RPS-32, six-vehicle Rail Grinder number DR79226 built in Italy, arrived at the RTC Derby for acceptance testing before being put to work on the UK rail network. The statutory static tests were followed by self-powered ride and brake stopping distance performance tests carried out between Derby & Bedford; and as with the earlier train, operating successfully at speeds up to 60 mile/h. The test instrumentation consisted of a portable battery powered, tri-axis vibration measuring unit incorporating a thermal strip chart recorder, that was used to assess the ride performance on-board the train. A new design of portable deceleration meter was used during the brake testing that could derive the brake stopping distance without the need to install a separate tachometer or doppler radar unit to the vehicle under test.

Further rail grinding machines were obtained during the 1990s including another self-propelled unit, a type SPML 15 three vehicle machine number DR79200. This machine was built by Loram in Minnesota USA and arrived for testing in 1994. Static tests included bogie rotational resistance, braking system and parking brake tests; and a self-powered ride and braking test was also carried out on the Midland Main line.

Fig.160 - Speno RPS-32 rail grinding train DR79226 at the RTC Derby, February 1990

RAIL CRANES

The 76 tonne Diesel Mechanical Breakdown Crane ADRC96200 was built in 1964 by Cowans, Sheldon & Co. Ltd of Carlisle, and during its refurbishment in 1987 the crane was fitted with air brakes, having previously only been fitted with a vacuum braking system. The modifications as a result of fitting the new air braking system, meant that tests were required to confirm the acceptable brake stopping distance performance of the crane, including its associated jib carrier wagon. Static tests were carried out at the RTC Derby including brake pressure measurements, brake cylinder application and release timings, and a handbrake performance test. Slip/brake tests were then undertaken on 28th November 1987, the test train being unusually hauled between Derby and Crewe using a single Class 20 locomotive, 20097, running nose end first. After arrival at Crewe, the Class 20 was replaced with a Class 47, 47113 for the purposes of slip/brake testing between Crewe and Winsford.

Fig.161 - ADRC96200 76t crane at Winsford after slip/brake testing, November 1987

When slip/brake testing, it was always important to establish the brake performance was satisfactory at lower speeds before increasing to the maximum designed operating speed. With the crane this became very apparent as we increased the slip/brake test speed because the brake stopping performance was somewhat worse than anticipated and at speeds above 50 mile/h. The deceleration rate with the brake applied was initially very low until the speed had reduced below about 30 mile/h, when the effectiveness of the cast iron block brakes increased. When a vehicle under test exhibited a stopping characteristic such as this the Officer in Charge allowed the locomotive and test car to coast further before instructing the train crew to stop in order to maintain a safe distance from the slip vehicle.

On occasions the DM&EE test team were called upon to undertake tests as part of a derailment investigation. Just such an instance following a derailment of a Regional Civil Engineers department (RCE), Heavy Duty Diesel Electric Crane, resulted in DRF81312 built in the late 1950s, arriving for static and dynamic testing at RTC Derby in 1988 coupled with jib carrier bogie bolster 'C' wagon DB922648. Following recording of the wheel profiles and the standard torsional stiffness ($\Delta Q/Q$) & bogie rotational resistance tests, crane DRF81312 was coupled to Test Car 1 and instrumented to investigate lateral stability at speeds up to the then current operating speed of 35 mile/h. A dynamic test run was conducted on the 28th January 1988 between Derby and Corby which highlighted lateral instability (hunting) and subsequently resulted in a speed restriction of 30 mile/h being imposed for the operation of this type of crane.

Fig.162 - DRF81312, 12 tonne crane undergoing preparations for test, January 1988

Two new designs of rail mounted general purpose light duty cranes were built by Plasser & Theurer in 1986, the first being an EPV-360 fitted with excavator equipment LDRP96505 which arrived in April at the RTC Derby for test. The second was vehicle LDRP96513 of the GPC-38 type, 4 tonne capacity crane that was submitted for test in September 1986. Following completion of the normal wheel weighing, torsional stiffness ($\Delta Q/Q$) and bogie rotation tests the cranes were coupled to the Test Car with instrumentation fitted to enable riding properties and brake performance characteristics to be measured. The cranes were designed to be able to move at slow speeds under their own power, therefore in addition the traction performance was also tested by measuring the drawbar load whilst hauling a trailing load of a test coach and a locomotive.

Fig.163 - LDRP96505 EPV360 excavator at Corby, April 1986

On the 26th April 1986 riding tests for the EPV-360 Excavator were carried out on between Derby, Corby and Cricklewood with LDRP96505 sandwiched between Test Cars 1 and 2. During the test run there was a problem en-route at Corby, whereby the spare excavator bucket that was mounted on its storage pillar on the excavator vehicle became unsafe due to failure of its underframe storage pillar mounting brackets. The excavator bucket had to be removed from the vehicle onto the lineside outside Corby signal box so the test run could continue. The traction and braking performance of the EPV-360 Excavator was evaluated the following day at the Mickleover test line. The results of the tests were satisfactory to allow the EPV-360 vehicles to be hauled in trains at speeds up to 45 mile/h; however a retest was recommended after a period of operation in service.

Fig.164 - Plasser GPC-38 craneLDRP96513 at the RTC Derby, September 1986

On the 29th September the first GPC-38 crane tests were conducted at the Mickleover test line which highlighted a minor issue with the suspension vertical damping when traversing jointed track. The braking performance and traction performance were proven to be acceptable and within limits. On return to Derby the following day a series of wedge tests were undertaken to determine the level of damping available within the suspension by evaluating the rate at which the oscillation of the suspension decays after rolling the vehicle over the test wedge. Minor modifications to increase the damping levels were trialled during two further test runs, this time over the Midland Main line route between Derby and Corby on the 5th November and between Derby and Bedford on the 12th November 1986. After completion of the tests the results were analysed and it was determined that the vehicles could be allowed to operate in train formations at speeds up to 45 mile/h, and as with the excavator, a retest was recommended after a period of operation in service. Both types of crane were re-presented for repeat ride testing during 1989, EPV-360 Excavator number LDRP96504 & GPC-38 4 tonne Crane number LDRP96512 being tested between Derby and Corby within the same test train coupled to Test Car 1 on the 15th May 1989. The results of the tests concluded that both types of crane could continue to be hauled in trains at speeds up to 45 mile/h.

In mid-1990 the GPC-38, 4t Crane, LDRP96513 was again back at the RTC Derby for evaluation of a modification to assess whether the operating speed of 45 mile/h could be increased. The crane was again instrumented to measure ride characteristics with additional suspension potentiometers and a test run planned between Derby and Crewe. The test run set off hauled by Class 31 locomotive 31420; however the locomotive subsequently failed at Egginton Junction due to running out of fuel. The test train was rescued quite promptly by two Class 20 locomotives 20170 + 20084 and returned to Derby. On inspection it was found that the Class 31 locomotive fuel tank level pipe was blocked which falsely showed the tank was full. A second attempt at the test run between Derby and Crewe on the 4th July 1989 was more successful although ride performance of the GPC-38 crane was not improved therefore the operating speed was maintained at 45 mile/h as confirmed during the initial tests in 1986.

ROAD - RAIL VEHICLES

A number of road/rail prototype vehicles were presented to the DM&EE between 1985 and 1993 for assessment and acceptance testing. The various vehicle types were designed for very different uses including personnel carriers, general engineering use and recovery support. The vehicles on which I worked, either leading or supporting the static and dynamic testing were:-

- Bedford Crew Cab Recovery Van, BRUFF conversion registration B155VRV

- Bedford Crew Cab Recovery Van, BRUFF conversion registration B445WPO

- Land Rover long wheelbase pick-up, BRUFF conversion registration D695AVT

- Mercedes Personnel Carrier, BRUFF conversion registration E192JAB

- Permaquip / Volkswagen Multicar 3.5 tonne pick-up, registration F511LRR

- Permaquip / Volkswagen Personnel carrier L771NAU

The Bedford Crew Cab Recovery Van, BRUFF B155VRV was presented for test in early 1985, commencing with static testing undertaken at the RTC Derby in the EDU workshop and weighbridge, with operational checks of the road/rail system, confirming the process for getting the vehicles on and off the track safely. We then measured the wheel weights and undertook a torsional stiffness ($\Delta Q/Q$) test taking particular note of the weight distribution on the rail wheels and the road wheels to ensure that the right amount of load was carried. Too little load on the rail guidance wheels and the potential for derailment increases, and too little load on the road wheels reduces the traction available to move the vehicle along the line.

Fig.165 - BRUFF Bedford crew cab recovery van at Mickleover, August 1985

Initial tests with B155VRV revealed minor issues with the rail guidance wheels hydraulic system and following modifications a second BRUFF Recovery Van, registration B445WPO was presented in August 1985 for test, following which the wheel weights and torsional stiffness ($\Delta Q/Q$) tests were repeated. Ride and braking tests to ensure safe operation was carried out on-track at the 8 mile long British Rail Research operated test line between Egginton Junction and Mickleover. Accelerometers were used to determine riding properties & vibration levels, and potentiometers provided assurance of the movements of the suspension and rail guidance equipment. In the case of the Bedford BRUFF, additional tests were required to support validation of the modifications applied to B445WPO, including dynamic wheel loads and structural integrity of the rail guidance equipment. Strain gauges were installed on the mountings of the hydraulic rams to measure the stress levels and dynamic forces, whilst the system was being operated to transfer the vehicle from road to rail mode and also when running along the Mickleover test line.

Fig.166 - BRUFF Bedford crew cab recovery van on the test track, August 1985

The crew cab became home for a temporary test set-up with signal conditioning amplifiers, a data recorder and a paper trace chart recorder. To power the test equipment a small portable generator unit was installed in the rear of the BRUFF. At the time the Bedford Crew Cab Recovery Van conversion by BRUFF was quite a unique vehicle being fitted with its own hydraulically operated lifting turntable which enabled it to be turned without the need for a flat hard standing area.

The pictures taken at the British Rail Mickleover test track show how the system worked in practice. This system made it very easy for us to undertake the tests on the vehicle. Once we had driven the five miles from the RTC Derby to the base for the test track at the old station yard at Mickleover, the generator was started and after the instrumentation had warmed up the BRUFF was easily put on-track on the hard standing area.

Fig.167 - BRUFF Bedford crew cab recovery van, rail gear instrumentation

Whilst the hydraulic rail guidance system was being operated during the on-tracking process, the measurements of stress levels in the hydraulic rams mountings was undertaken, and the potentiometers were then connected between the vehicle underframe and the rail guidance axle. These road/rail vehicles cannot be operated on the railway network during normal trains operations; only being allowed on-track after stopping all train movements in the area and possession of the line being arranged with the signaller. The Mickleover test line provided an ideal section of railway for us to undertake the tests on the BRUFF, this was pre-booked by arrangement with the British Rail Research department who operated the facility at the time and we had sole use on for the day of the whole length of the test line from Mickleover to Egginton Junction. The test line contained many track features eminently suited to the testing of road/rail vehicles including jointed and welded rail sections, concrete, wooden sleeper sections, a paved track section, points and crossings, various rail types including 85 & 95lb/ft. bullhead rail and 109 & 113lb/ft. flat-bottom rail, 30 & 60 chain radius curves and gradients up to 1-in-110. The test runs started with low speed movement along the line, assessing the measurements of the acceleration levels, suspension movements, wheel loads and stress levels on the chart recorder, in the rather cramped (and very warm) crew cab, providing us with assurance that it was safe to increment the speed slowly. The running behaviour of these road/rail vehicles was affected much more by discrete track features such as dipped rail joints and points/crossings, therefore careful assessment at different speeds traversing these track features was necessary. Following the Mickleover tests, the BRUFF was driven across to the Plant & Machinery Central Training Site at Toton where further tests were conducted driving the vehicle on, and across tracks, and traversing different types of points and crossings. Analysis of the test results was carried out back at the RTC Derby to evaluate the stress levels in the rail guidance system and the running characteristics. The BRUFF test vehicle B445WPO was allocated to the Western Region after completion of the tests and (at the time of writing) is still in existence in preservation at the Plym Valley Railway.

A new vehicle conversion produced by Permaquip was presented for test in August 1989, being a Volkswagen based Multicar Pick-up, (registration F511LRR). This vehicle was quite small and light-weight compared to the BRUFF Bedford vehicle, it had wide tyres that made it possible (with practice and on a dry day) to transfer the vehicle onto the track without the need for a flat hard standing area. The rail guidance wheels were fitted with a resilient inserts to aid noise reduction when operating on-track. After completing the standard torsional stiffness ($\Delta Q/Q$) tests and weighing at Derby, the vehicle was driven by road to the Annesley Colliery line in Nottinghamshire on the 6th September where I had arranged sole access to a few miles of line to undertake the running tests. The Permaquip vehicle types did not have any method of turning the vehicle therefore in order to change direction, although the vehicle could be driven in reverse at slow speed, to speed up the testing we had to de-track to turn the vehicle, then put it back onto the rails again.

Fig.168 - Permaquip/VW road rail Multicar at Toton, September 1989

Due to the small size of the vehicle cab there was only room for myself and the driver during testing, with a portable acceleration data recorder wedged between my feet I managed to collect sufficient data in numerous runs up and down the line to establish safe operating speeds of 35 mile/h on plain line and 15 mile/h when traversing points and crossings. The braking performance of these light-weight road/rail vehicles was extremely good, however this could be affected significantly by the rail head condition and road wheel tyre pressure. Some additional tests on this vehicle were also conducted at the Plant & Machinery Central Training Site at Toton; these included further tests to establish the capability for on/off tracking with different track configurations and varying levels of ballast.

The ride, stability and braking testing of another Permaquip vehicle, this time a Volkswagen Personnel carrier (registration L771NAU) were undertaken during August 1993 initially at the Derby RTC then on the Midland Railway infrastructure starting at the Swanwick Junction site of the preserved railway line. The vehicle operation was assessed to be acceptable up to speeds of 35 mile/h when on-track, however when traversing points and crossings the speed had to be reduced to 15 mile/h to maintain safe levels of stability. It is believed that this vehicle remained in the Derby area and was noted to be painted in Balfour Beatty colours about ten years after the testing had been completed.

Fig.169 - Bruff Landrover at Toton, September 1989

The Land Rover pick-up conversion by BRUFF (registration D695AVT) was presented for test at the same time as the Permaquip pick-up in August 1989. The design of the rail guidance system fitted to the Land Rover differed from that of the Permaquip having smaller dual guide wheels at each corner rather than a single larger wheel arrangement. Following the static testing and set-up checks at the RTC in Derby, the Land Rover underwent running tests on-track at the

Mickleover test line and also further slow speed tests at the Plant & Machinery Central Training Site at Toton. The test format followed must the same pattern as for the Permaquip vehicle resulting in an acceptable operating speed of 35 mile/h when traversing plain line and reducing to 15 mile/h over points and crossings. With the greater number of rail guidance wheels, and no space in the wheels to have a resilient insert, the Land Rover was noted to generate a bit more wheel/rail noise in operation on-track compared to the single wheel rail guidance system design.

TESTING OF PASSENGER ROLLING STOCK

DIESEL MULTIPLE UNITS

The Class 154 Diesel Multiple Unit (DMU) was initially a 2 car formation unit numbered 154001, which was used to prove some of the traction systems and a prototype of the air conditioning system being considered for the forthcoming Class 158 units that were at the time being designed. The two driving motor cars 55201 and 55301 were originally formed in the 1984 built as Class 150 prototype unit 150002, and underwent modifications starting in late 1986 with the fitting of Cummins NT855 diesel engines and Voith Hydraulic T211r transmission to car 55201, whilst 55301 had a modified version of the Twin Disc transmission that was originally tested in a Class 151 unit in 1985. The designed operating speed of the Class 150 units was 75 mile/h however the new Class 158 units were to be a 90 mile/h unit, therefore tests were conducted on 154001 up to 90 mile/h.

Fig.170 - Class 154 at Buxton fuel point, February 1987

Preparations for the testing commenced in early 1987 at the RTC Derby, with the interior of one vehicle being fitted out with instrumentation racking and tables to accommodate the signal conditioning and data recording equipment. The tests to be undertaken included vibration monitoring, interior noise levels, interior heating and cooling temperature levels, traction performance, braking performance, various engine temperatures and pressure measurements, and fuel consumption. I had the opportunity to join the passenger group test team to support them with a number of the tests on the Class 154 unit, firstly with the two car unit loaded to crush condition using bags of ballast, and undertaking braking tests on the Midland Main line route between Trent Junction and Leicester on the 4th February. The following day the planned test route took us from Derby to Birmingham then onto Bromsgrove, subsequently completing a number of test runs up the 1-in-37 gradient of Lickey incline between Bromsgrove and Barnt Green. A number of the traction performance tests were completed with each one of the motor cars isolated in turn, including standing start tests on the steep gradient. On the 6th February we headed north from Derby to Manchester and then onto Buxton for fuelling before completing two return trips between Buxton and Hazel Grove, again undertaking traction performance tests. During one of the test runs use was made of the curvature of platform 12 at Manchester Piccadilly to provide better access for an engineer to attend to a test instrumentation problem on the engine of car 55201.

Fig.171 - 154001 at platform 12 in Manchester Piccadilly, February 1987

The content of the DMU traction and engine parameter testing was different to that of my normal work at the time on freight vehicles; however the instrumentation including transducers, signal conditioning and data recording equipment used was very much the same. Further traction and braking tests were undertaken on the 11th and 13th February between Birmingham, Bromsgrove and Derby to Cricklewood via Manton before some of the test weights were removed from the Class 154 to then represent the fully seated load condition. I was involved in one more Class 154 test to Birmingham and the Lickey incline on the 20th February before returning to the freight test group for the upcoming SPA/OCA wagon bearing failure investigations project. The two car Class 154 unit was subsequently released into passenger service, returning on occasions during May and June 1987 for assessment of bogie stability, the unit was subsequently noted on the 10th June on a test run to Cricklewood. In September 1987 the unit returned to the RTC again, this time for re-formation as a 3 car unit with the inclusion of the intermediate motor car 55401, which had by then been modified for 90 mile/h operation. Prior to being returned into passenger service the 3 car unit underwent braking tests on the 29th September 1987. The Class 154 was seen again at the RTC Derby between mid-March and early April 1989 following modifications to the braking system during which time both 2 and 3 car formations braking tests were carried out with the unit in both tare and crush load conditions on the Midland Main line between Derby and Leicester. In addition on the 6th April 1989 the 3 car Class 154 unit was operated on a test run over the Buxton to Hazel Grove route to monitor engine cooling system performance.

ELECTRIC MULTIPLE UNITS

The Class 455 units were built in batches between 1982 and 1985, the later units being formed of three new vehicles and one existing (ex-Class 508) trailer car. During January 1988 the DM&EE testing team were called upon to assess the ride comfort on the later batch Class 455/7, including monitoring of the ride comfort in the Class 508 trailer car vehicles. The portable test equipment consisting of a tri-axial accelerometer/signal conditioning unit powered by a 12 volt battery, and a TEAC 8 channel data cassette tape recorder was prepared in the test section instrument room at the RTC Derby. In order to ensure correct operation of the portable instrumentation set-up, and validate the analysis process before embarking on the Class 455 test runs, a number of short proving runs were undertaken on Class 150 DMUs operating in normal service between Derby and Matlock. The data from the cassette tapes was then checked in Derby before Peter Metcalf and myself headed south to Orpington to get aboard the Class 455 test unit 5701.

Fig.172 - Test Unit 5701 after arrival at Ashford platform 2, January 1988

The tri-axial accelerometer unit had adjustable pointed feet to allow the unit to be levelled on the saloon floor before commencing the test run; the 12 volt battery being secured to the top of the accelerometer unit to stop it from moving during the tests. An analogue data output from the accelerometer/signal conditioning unit was fed directly to the TEAC data recorder which used high quality cassette format tapes that could capture approximately 1.5 hours of data with this particular set-up. This was just sufficient to collect data on a complete Orpington to Ashford test run on one cassette; but meant there was the need to change the cassette before the start of each test run.

Fig.173 - Portable ride comfort test equipment Class 455

The accelerometer/signal conditioning unit included a filter circuit to ensure that the frequency content of the analogue data was suitable for post-test analysis and did not include higher vibration frequencies (above 100Hz) or signal noise. Limiting the frequency content of the data also ensured that the data was captured correctly, without the risk of aliasing, onto the TEAC recorder which had a limited bandwidth for recording of around 1000Hz per channel. In order to undertake a full assessment of ride comfort levels for passengers that are seated, measurements should ideally be taken in the vertical, lateral and longitudinal planes at floor level, and at the base of the seat and on the seat back. For standing passengers the three planes are measured just at the floor level and in the case of these Class 455 unit tests, due to the need to gather data quickly and efficiently using the portable equipment and to enable comparison with other previous test recordings taken on other unit types, the levels of comfort were only evaluated at the floor level.

On the 29th January 1988 a number of dedicated test run trips between Orpington and Ashford (Kent) were undertaken at speeds up to 75 mile/h to obtain ride comfort data inside the saloon areas of the unit whilst operating over the routes simulating the running pattern of a normal service train.

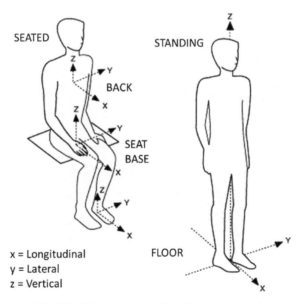

Fig.174 - Measurement positions for passenger comfort

The data cassettes from the Class 455 tests runs were returned to the analysis laboratory in Derby for post-test analysis. The method of analysis in this case was based on the requirements of a European standard using ISO2631 methodology and filter characteristics to evaluate the effect on the comfort levels experienced by passengers. The acceleration data in each plane, vertical [z], lateral [y] and longitudinal [x] was filtered and processed to only retain the acceleration levels at frequencies that make us feel most uncomfortable. The frequency range of accelerations expected to impact ride comfort in rail vehicles includes 0.5Hz to 20Hz in the vertical plane, 0.5Hz to 10Hz in the lateral and longitudinal planes and 0.1Hz to 2Hz during transitions into curves. In the case of the Class 455 units the levels of comfort evaluated from the data recorded in all vehicles during the test runs between Orpington and Ashford was demonstrated to be comparable with other newer generation EMU types, and also comparable between each of the vehicles on the unit tested when operating at speeds up to 75 mile/h.

The Advanced Suburban Bogie was developed by the British Rail Research department, initially being fitted and tested under an ex-Class 210 trailer coach that was designated Laboratory 8. The project was developed further in 1994 by the fitting of the bogies under a Class 466 EMU that was presented to the DM&EE for acceptance testing; it was this part of the project that gave me another chance to support the passenger rolling stock test group. I lead the project to undertake the statutory static tests at the RTC Derby, further sway tests at Strawberry Hill depot, plus dynamic ride tests on the Midland Main line that included stress measurements of the new design of inboard axlebox bogies.

Fig.175 - ASB - Class 466 at RTC Derby, April 1994

Class 466 unit 466028 comprising of DTOS vehicle 79339 and DMOS vehicle 64887 was delivered to the RTC Derby in late March 1994 and testing commenced quickly with the completion of bogie rotational resistance, torsional stiffness ($\Delta Q/Q$) tests and a parking brake test. We then started the preparations for the body sway testing which included the manufacturing of brackets to mount the prism reflectors to the bogie frames and wheelsets. Sway testing is a method of supporting the verification of the Kinematic Envelope (KE) which is the outline of the space occupied by a rail vehicle when in motion, including the effects of suspension movements, body roll and body sway. The prism reflectors were also mounted on the end of the vehicle body at various positions, being used as targets for taking the sway test measurements using a theodolite to establish the exact position of the wheelsets, bogie frame and body at each lift point during the test.

Fig.176 - ASB - Class 466028 at the RTC Derby, April 1994

The sway tests for passenger stock fitted with air suspension systems were conducted with the air suspension in the normally inflated condition and also in the deflated state. For the inflated condition tests the air suspension had to be maintained at its normal working height throughout the tests, therefore it was necessary to disconnect the suspension levelling valves and seal-off the air feed and cross equalisation pipes to prevent the suspension system from trying to level the vehicle as the test progressed. This often provided quite a challenge for the test team because it was necessary to maintain the air within the system, with no leaks for the duration of the test, which in some cases could be as much as 12 hours. My diary logs for the Class 466 sway tests note that preparation took a number of days to get the air system leak free to enable a sway test to be completed without leaking, and on one occasion having to repeat a test because the suspension had deflated before the end of the test. The DTOS vehicle of the Class 466 fitted with ASB bogies was sway tested on the 7th April 1994 at the RTC Derby, and the DMOS vehicle sway tests were carried out at Strawberry Hill depot on the 25th and 26th May. More detail about the method of sway testing is included in a later chapter of this book. The dynamic testing of the ASB commenced with the ride performance tests including the measurements of bogie strains being carried out between Derby and Leicester on the 11th May 1994, and also with the air suspension deflated between Derby and Kettering on the 12th May 1994; on both occasions hauled by a Class 47 locomotive with the Class 466 coupled between barrier coaches.

The Class 310 Electric Multiple Units were built in Derby from 1963 as 4 car units, originally classified AM10 to work on the newly electrified West Coast Main line routes.

Fig.177 - 310102 in preparation for brake testing at Bletchley, June 1995

During 1995 the operational requirements resulted in some of the units being reduced from a four to a three vehicle formation with the removal of the intermediate trailer coach. The change in formation meant that the braking performance had to be verified to still be within limits before the units could be operated as a 3 car in passenger service. I was allocated the responsibility to arrange and undertake the verification work that included the measurements of static brake characteristics of the 3 car unit, fitting the test instrumentation to accurately record the brake stopping distance performance, and executing the tests on the first of the modified units, number 310102. Preparations for the testing on 310102 were undertaken at Bletchley depot on the 17th June 1995 followed by brake stopping distance tests at various locations whilst operating up to 75 mile/h between Bletchley, Stafford and Crewe over a 5 day period. The first two test days being the 18th and 19th June with an empty unit, following which loading of the 3 cars with a large quantity of 20kg sand bags was carried out to simulate a maximum load of passengers. Loading of the Class 310 was a tiring task with nearly 200 seats in a 3 car unit that needed 80kg per seat of weight, making a total of about 16 tonnes of sand bags to be moved by hand into the unit. The design of the Class 310 unit with multiple bodyside doors did however assist the loading because the sand bags did not have to be carried very far once inside the unit.

Fig.178 - Loading of the Class 310 driving trailer to simulate passengers

The final two days testing on the 24th and 26th June again between Bletchley, Stafford and Crewe concluded the measurements of the brake stopping distance performance in the fully loaded condition. Once the testing was completed the sand bags load had to be removed, and the whole unit interior cleaned prior to handing back to the operations department. The acceptable results of the braking tests were plotted in the standard format and a short report completed for approval prior to commencement of service operation of the reduced formation units shortly after the tests.

LOCOMOTIVE HAULED COACHING STOCK

In October 1987, a pair of prototype BT41 bogies, (made by SIG Swiss Industrial Company) were delivered to the RTC Derby along with Mk III TS coach 12140 to trial the new bogies which were destined for fitting under the new Inter City 225 (IC225) service Mk IV coaches that were still under construction at the time. The coach was selected for trials having been previously used for a bogie trial with an early version of the BRE T4 bogie during late 1985. Minor modifications to the coach to accept the new SIG bogies were completed at the EDU, and the bogies fitted. Installation of ballast weights to simulate the new Mk IV coach body weight was necessary before the test team could commence weighing and static tests to establish the torsional stiffness ($\Delta Q/Q$) and bogie rotation characteristics. The TS coach 12140 was one of the London Midland region loco-hauled type Mk III coaches, therefore also required fitting with the HST type 36-way jumper cables and connections to enable operation within the proposed test train formation using Class 43 High Speed Train (HST) power cars. Although Test Car 10 was ideal to support this project, it was being used for another project at the time; therefore the base for the test instrumentation and engineers during the test programme was a borrowed TGS vehicle 44101. In order to house the amount of test instrumentation inside the TGS, a number of rows of seats were removed and temporary instrument racking and tables were installed.

Fig.179 - Test installation in TGS 44101, October 1987

The 240 volt electrical power for the test instrumentation was provided by a portable diesel generator unit which was installed in the luggage area of the TGS vehicle. Inside the TGS the instrumentation racking housed the signal conditioning and amplifier units that conditioned the raw data signals from the transducers on the coach, providing analogue signal voltages that were recorded at suitable levels onto magnetic tape recorders and ultraviolet paper trace chart recorders. The amplifier and signal conditioning units were designed by the DM&EE Testing Section development group specifically for the purpose of on-train data collection, they were housed in standard 19" instrument cases of varying depths to accommodate stabilised power supplies, filtering, electrical and shunt resistor calibration functions. In order to provide a comparison between the BT41 bogie fitted coach 12140 and a standard HST coach, TO vehicle 42317 was also provided for the tests. The instrumentation transducers fitted to coach 12140 and the BT10 bogie fitted coach 42317 consisted of accelerometers and potentiometers to monitor the running characteristics, body vibrations, suspension movements and passenger ride comfort levels. In addition strain gauges were installed in specific areas of one BT41 bogie frame to record stress levels during the dynamic running for evaluation of the structural integrity of the new design of bogie operating on the UK infrastructure. This project was also the first major DM&EE test project to use an on-board analysis computer system to provide the engineers with analysed test results on the test train, rather than having to return the data to the analysis laboratory back in Derby for post-test analysis. The on-board analysis was used to determine passenger ride comfort levels in the form of a Ride Index. The computer system used the then up to date HP Basic type 9816 computer connected via a HPIB interface to a Microlink real-time data sampling system which digitised the filtered and amplified analogue acceleration data signals; then counted the frequency of acceleration, the number of cycles and the amplitude of the acceleration levels, finally sending the analysed data to the HP computer for storage on a state of the art 40MB Winchester Hard Disc drive. The one-inch reel to reel magnetic tape recorders used were of the 42-track type, which provided ample channel capacity for all the data channels. The ultraviolet paper trace chart recorders were capable of handling up to 36 channels, although about 15 channels were normally used in order to give a readable trace output without too much channel overlapping. Following the completion of the instrumentation fitting and the calibration set-up, the three coupled coaches (42317, 12140 and 44101) were moved within the RTC sidings with the Class 08 shunt locomotive for preliminary checking of the instrumentation channels functionality. The three coaches remained coupled together from the start of the instrumentation fitting because the majority of the 40 or so transducer cables passed through the gangway between the vehicles. On the 28th October the two Class 43 power cars (43102 and 43155) and a further HST trailer vehicle (44090) arrived at the RTC in Derby. Additional cables were added to connect the independent intercom systems and train test speed display units between the TGS, and the Class 43 power car cabs.

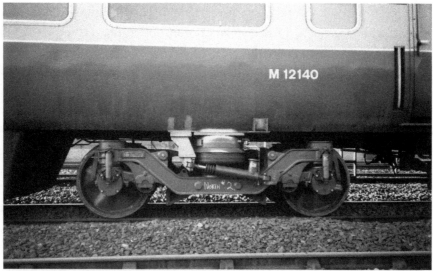

Fig.180 - Mk III TS coach M12140 prototype BT41 bogie, October 1987

The whole formation was coupled together on road number 3 at the EDU workshop with the power cars sticking out both ends of the workshop which meant the workshop doors had to remain open. It did however allow the power car engines to be run without filling the workshop full of exhaust fumes; there being no exhaust extraction equipment fitted within the EDU workshop at the time. We completed the installation of the independent cab intercom system and checking of the 36-way HST multiple working connections throughout the train.

Fig.181 - HST power car 43155 at the RTC Derby in preparation for test, October 1987

Before commencing dynamic tests the SIG bogie fitted coach 12140 and BT10 coach 42317 were also both loaded with sandbags to simulate a full passenger load. Plastic cover protection was fitted to the seats and four of the 20 kg sandbags were placed on each seat. Additional bags were placed in the luggage racks at the end of the coach to simulate standing passengers and luggage. The test running commenced on the 29th October 1987, with a short instrument proving run between Derby and Kettering before running up to Heaton Depot just north of Newcastle on the 30th October; this became our base for the next two weeks for the series of tests on the East Coast Main Line between York and Darlington and also between Newcastle and Edinburgh. On arrival at Heaton the north end power car 43155 was swapped for 43159, and the testing with the 2 power cars plus 4 coach formation commenced in earnest on 31st October 1987 between York and Darlington. The initial runs were at slow speed to allow the collection of at least 10 miles of running characteristics data for the new bogies at each test speed; then incrementing the test speeds up run by run in 5 mile/h increments until we reached a maximum test speed of 125 mile/h on the first day. The running characteristics of the new BT41 bogies were recorded and assessed during, and at the end of each test run being analysed by manually inspecting the chart time history traces to ensure all was in order and it was safe to increment the speed higher. The test programme and operating arrangements were set up to simulate running speed profiles aligned with a typical HST speed profile and also the new 140 mile/h speed profile proposed for the IC225 trains.

The 1st November 1987 saw further incremental speed test running between York and Darlington, taking speed up in 5 mile/h steps above 125 mile/h to confirm the stability up to 145 mile/h. In fact during one run the speed attained was slightly higher, and we achieved what became a diesel speed record of 148 mile/h over a measured mile on a southbound run between Darlington and York, with an absolute maximum recorded of 148.5 mile/h just south of Northallerton. The maximum speed attained was not constrained by the running characteristics of the SIG bogies under test, but by the fact that we had reached the balancing speed of the train formation with both Class 43s at full power within the operational constraints of the York to Darlington route section. The recorded 148 mile/h was accepted into the Guinness World Records and stands today as the fastest diesel-powered train in the world.

Fig.182 - HST power car with the SIG bogie test train at Darlington, October 1987

Further high speed running in line with the 140 mile/h speed profile continued on the 2nd and 3rd November running south to Bounds Green then returning back north to Edinburgh before stabling at Heaton, although we did not reach the speeds attained the previous day. On the evening of the 3rd November the sandbags were removed from the test coaches to revert to the tare condition; testing then continued until the 6th November on the East Coast Main line including runs between York, Bounds Breen, Newcastle and Edinburgh, again checking the stability up to 145 mile/h. My diary records show that due to brake disc problems with 43159 after the very high speed running another power car swap was made at Heaton during the week of testing, being replaced with 43104 on the 3rd November. The sandbags were again loaded into both test coaches at Heaton in the evening of the 6th November and a final laden condition 140 mile/h speed profile test run completed running south to Bounds Green then returning north to York on the 7th November.

Following the completion of the test running on the 7th November at York, the train was moved to Neville Hill where the power cars 43102 and 43104 and one of the HST trailer cars were removed. The remaining three coach formation was by this time very dirty, because we could not allow carriage washing due to the risk of damage to the test instrumentation. Class 43 power car 43008 was attached and hauled the three coaches from Neville Hill to Derby with a HST barrier coach coupled at the rear. Upon arrival back at Derby the Mk III trailer vehicle and the HST barrier were detached and returned north to Neville Hill; we then set about removing the test equipment from 44101 and 42317 so they could be released back to traffic. The results from the tests with the first SIG bogies fitted under coach 12140 proved very successful however to further improvements in the riding comfort levels were envisaged.

Fig.183 - 43008 at Neville Hill, November 1987

In mid-1988 following detailed evaluation of the results some minor design changes and modifications were made to the suspension of the SIG bogies before they were reinstated under 12140 for a further series of tests; this time operating on the West Coast Main Line (WCML). The standard Mk III comparator vehicle provided for these tests was number 17174, which was available at the RTC Derby having just completed a series of WSP system tests. This time we used Test Car 6 for the test equipment installation base, and additional brake power was provided by a Mk III sleeper coach number 10706, and Laboratory Coach number 23 (RDB975547) on loan from the British Rail Research Division. The majority of the test instrumentation had to be re-installed onto coach 12140 after the bogie modifications, and included additional equipment to measure the air suspension air consumption and flow rates. These series of tests commenced on the 25th July 1988 with an instrumentation proving run between Derby and Leicester, hauled by 47561, then returning back to the RTC Derby. The following day the train moved to Crewe, again hauled by 47561 where a locomotive change was effected and 87023 coupled on for testing on the West Coast Main Line between Crewe and Carlisle; returning south the same day to Crewe where the train was stabled at the electric depot overnight. On the 27th July the test was run again between Crewe and Carlisle hauled by Class 81 locomotive 81007, returning to Crewe where the locomotive was changed for a Class 47 for the return transit movement back to Derby. On the 28th July the train ventured out again, this time hauled by 47521 between Derby and Crewe, then running south on the West Coast Main

Line to Willesden and back. Data analysis from these test runs helped the SIG engineers evaluate a final minor modification to the suspension on 12140, and two further test runs were carried out on the 2nd and 3rd August between Crewe and Carlisle; in the same manner as the previous week, the train was stabled at the electric depot at Crewe overnight. Following completion of these tests the instrumentation was removed from the SIG bogies and coach 12140. The results of the Grayrigg to Lancaster section of these tests were most useful in providing the SIG engineers with the information they required to finalise the design of the bogies for installation under the first of the new Mk IV bodyshell that was nearing completion. The coach 12140 subsequently went on to be modified again and used to trial the in-service operation of the production version of the BRE T4 bogies on the West Coast Main Line. In January 1989 the first of the new Mk IV coaches arrived at the RTC Derby for testing.

Fig. 184 - Mk IV test train at Bounds Green, February 1989

Tourist Open End (TOE) Coach 12201 painted in the new Inter City livery was fitted with the production version of the BT41 bogies tested the previous year, however the interior of the vehicle at this time was as not fitted with any seats. The lack of seats meant that ballast bags had to be put inside the coach to represent the weight of the missing seats. The planned series of tests involved the use again of Class 43 (HST) power cars to provide traction, this meant that much like the tests with the Mk III coach 12140, the new Mk IV vehicle had to be fitted with a temporary 36-way cable and connectors at each end of the coach to allow the multiple working of the Class 43 power cars to operate in the normal manner at each end of the train. The preparation works started at the RTC Derby in late January 1989, commencing with static tests including weighing, bogie rotational resistance, torsional stiffness ($\Delta Q/Q$) and body sway testing. The body sway testing was undertaken to verify what is called the Kinematic Envelope (KE); this is the outline of the space occupied by a vehicle when travelling along the tracks. The sway test provided the engineers with actual suspension movements from simulated effects of the vehicle operating, by tilting the vehicle and measuring the suspension movements and body-sway angles. More detail about the method of sway testing is included in a later chapter of this book.

Fig.185 - Mk IV coach 12201, February 1989

Following completion of the static testing, the instrumentation set-up was started; for this series of tests things were made a bit easier because the Mk III test coach Test Car 10 was available for us to use, providing a readymade base of test tables, instrumentation racks, a powerful 3-Phase generator set that also powered the air conditioning system within the test car, and a well-appointed kitchen area with a microwave and a baby belling cooker. In order to provide brake power the test train had another Mk III vehicle marshalled in the formation so the new Mk IV TOE was sandwiched between Test Car 10 and a HST trailer coach TGS 44019 which was provided specially for the tests. The first run out on the network for dynamic testing was an instrument proving test run that was undertaken on the 8th February 1989 over the Midland Main line between Derby and Bedford hauled by Class 47 number 97545. The instrument proving was followed on the 9th February with further checks and incremental speed data collection at speeds up to 100 mile/h, during which data to demonstrate compliance with the dynamic resistance to derailment criteria was obtained. The Class 47 locomotive 97545 used for the initial tests had received a special maintenance safety exam prior to being authorised to operate up to 100 mile/h for the purpose of this test. The following day two Class 43 power cars arrived at the RTC Derby and were coupled onto the test formation, 43051 attached to Test Car 10 and 43116 attached to the TGS; checks were made to ensure the interconnections between the two power cars was working correctly and an independent intercom and speed system installed between Test Car 10 and the two driving cabs.

Testing on the East Coast Main Line route began on the 11th February following a transit movement between Derby and Grantham, we then undertook four return runs between Grantham and Connington Loop collecting data to demonstrate compliance with the dynamic resistance to derailment criteria up to 125 mile/h. One round trip was completed with the bogie yaw dampers removed to confirm the affect on the bogie stability in the unlikely event of a damper failure in service. On completion of the last run south the yaw dampers were re-fitted and the train continued onwards into Kings Cross before moving to Bounds Green depot where the train was stabled overnight. The East Coast Main Line tests were originally planned to operate up to the proposed IC225, 140 mile/h speed profile as per the previous series of tests in 1987, however operational restrictions prevented this and the whole programme was subsequently limited to 125 mile/h operation. Further tests continued until the 18th February 1989 including the tests with one bogie fitted with a cross-member design to simulate that which was later fitted to the BT41-C bogie derivative to be installed under the new Mk IV DVT vehicles. Once testing had completed on the 18th February, the whole train was moved back to the RTC Derby for decommissioning of the test installation.

Fig.186 - 43116 at Bounds Green with Mk IV test train, February 1989

Another series of BT41 bogie development tests were carried out on the East Coast Main Line during June and July 1989 following a number of modifications to the bogies including changes to damper rates, anti-roll-bar bush stiffness and bumpstop rubbers. Mk IV TO coach 12400 was supplied to the RTC Derby in early June 1989 for instrumentation fitting along with TOE vehicles 12200 &

12201, and TOD 12300. Test Car 10 was again used for the instrumentation base and coupled with the Mk IV coaches to form the train as follows, TC10 + 12200 + 12400 + 12300 + 12201. An instrument proving run at speeds up to 100 mile/h was operated on the 29th June 1989 over the Midland Main line route to Bedford. The following day the test train was hauled to Grantham by Class 47 number 47973, where a locomotive change was effected with a Class 91 to provide traction for two days of stability and braking testing on the East Coast Main Line. These tests were successfully operated up to 140 mile/h between Peterborough and Grantham on a number of runs on both days before returning to Derby in the evening of the 1st July hauled by 47973. The 3rd July saw the test train head north to Edinburgh stopping en-route at Neville Hill depot to swap the Class 47 for two Class 43 power cars that took the train forward for testing up to 125 mile/h on the northern part of the East Coast Main Line. The route between Newcastle and Edinburgh has a number of curves that were traversed at speeds where the cant deficiency reached the maximum allowable 6 degrees, in order to fully assess the ride performance of the Mk IV coaches. The train was stabled overnight at Heaton depot in Newcastle, and on the 4th July the two Class 43 power cars were again used to haul the test train south to Retford then returning north to Heaton via Edinburgh. The final three days of this series of tests began with the train heading south to Bounds Green depot hauled by a Class 47, followed by ride, stability and brake performance testing again up to 140 mile/h over a number of route sections between Kings Cross and Doncaster hauled by a Class 91 locomotive. The final day of testing on the 8th July was disrupted following closure of the East Coast Main Line due to a fire at New Barnet station; however some test runs were attained between Stevenage and Grantham eventually returning back to the RTC Derby in the late evening. Although these tests showed significant improvements in the riding and stability performance of the Mk IV coaches compared to the February 1989 tests, further modifications including various yaw and lateral suspension damper rates, bump stop types, and a bogie anti-pitch device were proposed in order to fine tune the ride comfort levels. A number of combinations of the dampers were prepared, which were colour coded for reference such that the changes could be easily applied between test runs. A total of fourteen days of test running were planned to start with six days set aside for the selection of the optimum yaw damper rate, then three days for lateral damper selection and two days for progressive lateral bump stop evaluation, finishing with final assessment over three days with the selected optimised set-up.

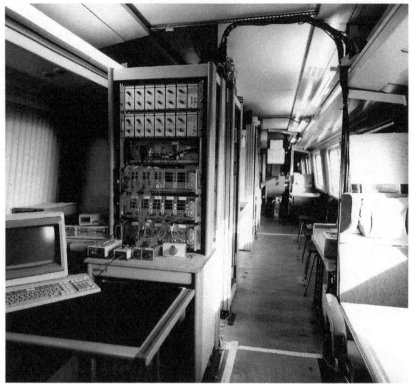

Fig.187 - Mk IV coach instrumentation area in 11200, October 1989

On the 1st November 1989 we commenced instrumentation installation and set-up at the RTC Derby on Mk IV TOD coach 12300 which was in the loaded condition, and using Mk IV Pullman Open (PO) coach 11200 for the test car base. A protective floor covering made from fire retardant plywood was installed in 11200 followed by a new design of instrument racks and tables to house the signal conditioning, data recording and the on-board analysis system. The power for the instrumentation systems was provided by a 240 volt diesel generator unit that was installed in the DVT vehicle 82204. My responsibility during these tests was for the installing and operating the ride comfort level assessment on-board analysis system (OBA), which was a further development of the original OBA system from 1987 but using the newer HP310 series basic computer coupled to an upgraded Microlink digitiser with data signal inputs from ISO filtered RMS to DC converted ride acceleration channels, train speed and route location. The OBA system computed the journey average ride comfort levels for twelve vertical, lateral and longitudinal acceleration data channels and provided printouts at the end of each test run of the results along with the attained speed profile for the test run. The speed profile traces were important, providing assurance that a sufficient length of the route had been covered with the train operating as near as possible to the defined test speed profile. Previously this had only been possible with post test analysis, so the benefits were realised quickly during this series of tests, where an assessment of the performance of each combination of damper settings was needed after each test such that decisions could be made with respect to the modification to be applied for the next test.

Fig.188 - Mk IV coach instrumentation area in 11200, October 1989

Testing commenced on the 17th November 1989 in the normal manner with a Class 47 hauled instrument proving run on the Midland Main line route between Derby and Bedford. Then on the 20th November 47973 hauled the test train from Derby to Bounds Green depot via Peterborough where 91008 was coupled in readiness for the start of the East Coast Main Line testing. Each day the test train was operated on return trips between Bounds Green depot and Grantham, running at speeds up to 125 mile/h south of Peterborough and up to 140 mile/h north of Peterborough on the section between Werrington Junction and Stoke Tunnel. In addition to the data analysis produced by the OBA system, hardware spectrum analysers were used to generate on-board frequency analysis from the acceleration and displacement data channels; providing supporting information to the engineers making the decisions for damper combination application for the next test day. Two different rates of primary vertical suspension dampers were tested, six types of yaw dampers, four lateral dampers and three types of lateral bump stops; with all the modifications work being carried out at Bounds Green depot between each of the test days. Most of the test days were completed without any problems, with Class 91 locomotive 91008 providing traction throughout and attaining up to two return trips to Grantham each day. Exceptions were on the 28th and 29th November when the DVT 82204 suffered a warning horn failure and also a signallers radio failure which in both cases resulted in an early finish for the respective day. The final day of testing was the 9th December 1989 and the last run was terminated at Peterborough where a Class 47 was attached to the test train and hauled the formation back via Leicester to the RTC Derby. The coaches were de-instrumented, cleaned, and the dampers removed from 12300 such that the characteristics of the dampers from the optimum set-up selected during the tests could be verified using the damper test rig in the EDU workshop.

The next stage of development for the Mk IV coaches included a modification to optimise the vertical ride comfort involving the rotation of the bogies, complete with the underframe mounting, through 180 degrees. This modification altered the location of the yaw damper mounting to be at the inner end of each bogie rather than the outer end, with the result that forces from bogie pitch movements were directed towards the vehicle centre rather than the vehicle ends. A repeat sway test was carried out on 12300 during January 1990 with the revised arrangement, with and without an anti-pitch device installed. These modifications were dynamically tested during another series of on-track tests during March and April 1990.

Fig.189 - Mk IV coach 12300 sway test with anti-pitch device, January 1990

The bogie modifications were applied to coach 12300, and for comparison two other Mk IV TO coaches 12400 and 12478 were included in the test train to which further lateral bump stop modifications and different rated lateral dampers were also applied during the tests. These modifications were carried out at the EDU workshop in Derby between each day of testing. I again used the OBA system and spectrum analysers to undertake on-board data analysis. By this time the OBA system had been developed further to automatically evaluate the ride comfort levels when traversing different track sections with varying levels of cant deficiency, which provided a greater level of detail for modification evaluation. The dynamic testing included a significant amount of running, totalling 28 days of operations that took the test train onto the Midland Main line to Bedford, the West Coast Main Line between Crewe and Carlisle and also the East Coast Main Line between Kings Cross, Doncaster and York. A wide variety of locomotives featured in hauling the test train during this series of Mk IV coach tests, including Class 31, Class 47, Class 87, Class 90 and Class 91. Following completion of the dynamic tests and before 12300 was released, a final series of static bogie rotation tests were carried out at the RTC; although I was not involved myself, having returned to freight vehicle testing work to look after the acceptance testing of a new ARC 102 tonne Low Track Force Bogie Hopper Wagon.

In March 1991 the Mk IV coach acceptance project again had my attention during the final series of tests following the installation of inter-vehicle dampers; these were located between the headstock and centre coupler bar at each end of each coach to further optimise the ride comfort levels. The test installation was limited to accelerometers mounted in three of the five coaches (12467, 12469 and 12468) to measure ride accelerations, and assess the effects of the application of the inter-vehicle damping on lateral stability, particularly when passing over points and crossings, and when traversing the more curvaceous sections of the East Coast Main Line route. The tests were carried out in two stages, firstly based out of Bounds Green depot for testing between Kings Cross and York at speeds up to 140 mile/h using a Class 91 locomotive; then moving to Heaton depot for testing between York and Edinburgh using two Class 43 power cars coupled back to back to haul the train. The use of the Class 43 power cars was necessary because the electrification was not yet completed to Edinburgh, and the back to back formation hauling the test train was necessary because the Mk IV coaches were not through wired with the requisite 36-way cabling to allow operation of the power cars at either end of the rake. Luckily powers cars 43080 and 43123 were available, being of the type modified for DVT use on the East Coast Main Line, therefore having standard buffers and drawgear fitted to the front ends. The outcome of the tests was successful confirming that the application of the inter-vehicle damping had further improved lateral ride comfort levels for the Mk IV coaches.

Fig.190 - BREL T4 bogie fitted under 10310 for trials on the ECML, July 1993

Although the Mk IV coaches were operating in service on the East Coast Main Line, I was again to be involved in testing with the vehicles in 1993 as part of the preparations for the introduction of the Nightstar sleeper vehicles. ABB Transportation installed a pair of T4-5 type bogies under a Mk IV Service Vehicle (SV) Coach number 10310 in order to evaluate performance of the bogie on UK infrastructure, and in accordance with the European criteria for running dynamics. A series of static and on-track tests were undertaken during July and August 1993 assessing the following elements in respect to the European criteria:-

- Stability

- Ride Quality

- Lateral Forces at Bogie Frame and Axlebox level

- Permissible Track Stress

- Safety against derailment over Twisted Track

- Negotiating Minimum Curve Radius

- Mass of the Bogie

- Maximum Axleload

- Tyre Profiles

- Operation with Deflated Air Spring

In addition to 10310, three more standard BT41 bogie fitted Mk IV vehicles were provided to the RTC Derby to support the testing; TOE 12219, TO 12528 and DVT 82229. Static testing commenced on 10310 with wheel weighing, torsional stiffness ($\Delta Q/Q$) and bogie rotation tests before conducting body sway tests. A temporary test installation was prepared in vehicle 12528 following the removal of a number of seats to allow a test equipment bench and instrument racks to be secured.

Fig.191 - Test instrumentation in 12528, July 1993

Accelerometers and potentiometers were fitted on both bogies on 10310, and strain gauges were installed to measure bogie frame stress and axlebox traction link stress levels. All transducers were connected with cables to the signal conditioning amplifiers mounted in the instrument racks in 12528. TEAC VHS type data tape recorders and Gould thermal paper chart recorders used to capture the data during testing. The power for the test instrumentation system was provided by a temporary 240 volt generator set installed in the DVT vehicle 82229. The wheel profiles installed on the T4-5 bogies of vehicle 10310 when delivered to the RTC were turned to a P8 profile, which was commonly used in the UK, and the same which was installed on the standard Mk IV BT41 bogies; however assessment of the running dynamics was also required with a P10 wheel profile, which is the UK equivalent of the S1002 profile used in Europe. The test programme was therefore set-up such that the wheels on vehicle 10310 could be re-profiled from P8 to P10 part way through the test programme. Dynamic test runs commenced on the 26th July 1993, hauled by a Class 47 locomotive making a number of test instrument proving, set-up and equipment calibration runs between Derby and Leicester. These runs continued on the 27th and 28th July with yaw damper mounting position optimisation runs, again between Derby and

Leicester. Once the optimum set-up for the bogie was achieved further test runs based out of the Derby RTC were carried out on the 29th and 30th July; on both days heading over to the West Coast Main Line and operating between Stoke on Trent, Bletchley, Nuneaton, Bletchley, Rugby and back to Derby via Stoke on Trent. On the 31st July we headed north setting out early from Derby behind 47976 via York and up the East Coast Main Line to Edinburgh, where the train was stabled overnight in Craigentinny depot. On the 1st August the train was taken to the wheel lathe at Craigentinny depot and the wheel profiles of the T4-5 bogies turned to a P10. The train was then returned into the depot to check the suspension heights, before loading with a total of 12.5 tonnes of ballast bags in order to simulate a full passenger load. Over the following two days the dynamic test runs were carried out down the West Coast Main Line via Carstairs, again hauled by a Class 47. On return to Craigentinny two more Mk IV coaches were added to the formation, and a Class 91 locomotive was provided for the remaining East Coast Main Line tests. The train formation was prepared as follows, 91022 + 12219 + 10329 + 10301 + 10310 + 12528 + 82229, and on the 4th August was operated south to York including testing at speeds up to 137.5 mile/h (125 mile/h + 10%) on the section between Northallerton and York to confirm stability.

Fig.192 - T4 bogie instrumentation, July 1993

The ballast bags were removed in the evening of the 4th August after return to Craigentinny, and the train successfully repeated the same high speed test run operations to York on the 5th August in the tare condition. The Class 91 locomotive and extra Mk IV coaches were then uncoupled and the test formation hauled by a Class 47 locomotive back to Derby, where the test instrumentation was removed and the coaches returned to a serviceable condition. Post test data analysis successfully demonstrated that the running dynamics of the T4-5 bogies complied with the UIC European criteria in place at the time.

LOCOMOTIVE TESTING

My first involvement with locomotive related testing came in 1984 when I had a brief interlude from the Class 37 locomotive project work I was undertaking with the DM&EE Freight Design team. On the 18th May I made my way south by service train to gather some data during a trial running test on the Western Region; for which two Class 56 locomotives were used to haul a 4600 tonne train of loaded PTA aggregate wagons for a performance test. I joined the train in the leading cab at Westbury Yard, departing at 1am the following morning. The train made its way via Newbury and Reading, arriving at Acton at about 5:30am. Following a train crew change at Acton we continued on to the unloading terminal at Purfleet, arriving about 7am. During this test run my task was to record as much information in a log report about the train run; this included, weather and rail head conditions, train speed, location (trackside mile posts), time, traction current, any wheelslip activity, and anything else that could be considered as useful information about the train operation. The result log report was presented to the Officer in Charge (OIC) upon my return from Derby, and used to support the evaluation of maximum train operating loads for freight traffic on different routes.

New Class 58 locomotive number 58001 arrived at the RTC Derby in early 1983 fresh from Doncaster Works for acceptance and performance trials, and I was lucky enough to start my apprenticeship placement with the DM&EE testing section just in time for the locomotives arrival. Static testing started straight away with wheel weight measurements undertaken on the EDU weighbridge on the 15th February, followed by a torsional stiffness ($\Delta Q/Q$) test and then a bogie rotation test on the specially designed test rig on road number 3 in the EDU workshop on the 16th February. This was my first experience of undertaking these statutory tests which were to feature in almost every future project I was to undertake over the next 17 years. We then began the preparation works with the locomotive coupled to Test Car 6, in readiness for the dynamic on-track testing. Before the dynamic acceptance tests could get under way, it was necessary to undertake a ride safety and stability test with the Class 58 hauled dead within a train. On the 18th February 1983 a Class 45 locomotive hauled the train formation of 58001 coupled to Test Car 6 and two passenger coaches between Derby and

Bedford and return operating at speeds up to 80 mile/h. The test team worked closely with the DM&EE design team in evaluating the various aspects of the new design of the Class 58 locomotive traction and braking performance, running characteristics, temperature levels and stress level assessments in a number of specific points on the underframe structure. This requirement for structural stress evaluation on 58001 provided me with an opportunity to learn the art of strain gauge installation which involved preparing the surface of the area to be evaluated, removing the paint down to bare metal and smoothing the surface of the metal to get an acceptable flat surface for sticking the foil backed strain gauge onto using special super strong adhesive. The idea being that as the metal bends, then the strain gauge also bends with it which causes the electrical resistance of the strain gauge to change. The strain gauge resistance and its properties were known, which meant its output could be calibrated to provide accurate strain measurements, which were used to evaluate the stress levels present in the particular location of the structure. Another chapter in this book provides more information from my experience of installing and taking measurements from strain gauges. A number of the strain gauges were located on the Class 58 in nice, easy to access positions on the underframe and bogie frames of the locomotive; however some of the locations that the design engineers were interested in evaluating were on the underframe cross members beneath the Class 58's Ruston Paxman 12RK3ACT power unit. The design of large bodyside doors of the Class 58 helped access; however many hours were spent lying on my back reaching under the power unit to prepare the metal surface of the underframe cross members and to install the strain gauges. During this time the weather was not being kind in Derby and the roof area on the EDU workshop sprung quite a bad leak, soaking the test team below, including myself who were busy installing the test instrumentation. For the majority of the Class 58 dynamic type acceptance testing, Test Car 6 was used as the base for the test team and housed the signal conditioning and data measurement equipment. Prior to commencing the dynamic performance testing, the Class 58 coupled with Test Car 6, was taken back to Doncaster Works and connected to the load bank test facility at the works to support the set-up and calibration of the equipment used to measure the engine output and performance characteristics.

A series of static tests were also conducted on locomotive number 58004 at Doncaster on the 2nd August 1983; including the measurements of exhaust gas emissions and noise levels; this involved various temporary frame arrangements to hold the test measuring equipment at the correct distance above the locomotive twin exhaust outlets.

Fig.193 - 58001 coupled to Test Car 6 at RTC Derby, May 1983 [Servo]

During the period between July and September the dynamic testing operations with 58001 progressed, predominantly on the Midland Main line because this was the route section over which the drivers trained on the Class 58 had their route knowledge. Traction performance running testing was undertaken in early August 1983 in two stages, firstly operating at speeds up to 60 mile/h with 58001 and Test Car 6 coupled to rakes of loaded HDA wagons of varying lengths operating between Toton and Cricklewood. A screw coupling fitted with a calibrated load cell was used between the Class 58 and Test Car 6 in order to measure the tractive effort exerted on the trailing load by the locomotive throughout the tests. At Cricklewood the Class 58, which was permanently coupled to Test Car 6 due to the test instrumentation cables, was detached from the HDA wagon train and moved via Cricklewood Junction and Dudding HIll Junction to turn, before re-coupling to the HDA wagons for the return trip. On return to Toton the 58001 plus Test Car 6 coupled formation was turned again, running via Attenborough Junction and Trent Junction in readiness for the next day of testing. The second stage of the traction tests was to assess the performance at speeds above 60 mile/h, for which a rake of eight Mk I coaches were used on the test runs between Derby and Cricklewood. In order to make the run-round process more efficient at Cricklewood, a Class 25 locomotive was attached to the rear of the formation to haul the train during the turning move which was carried out between Dudding Hill Junction and Cricklewood Junction. After the return run to Trent Junction, the Class 25 was used again to haul the formation back to Derby such that it was the correct way round for the following day of testing. Later in

August the strain gauging work on the bogies and axleboxes had been completed, and the Class 58 still coupled to Test Car 6 was moved into the Litchurch Lane works, to undertake curving tests on a tight curve of approximately 4 chains (80 metres) radius that was located within the works site. The purpose of the curving tests was to establish that sufficient lateral movement and forces at the centre axle on each bogie were suitable for negotiating tight curves. Further curving tests were undertaken before the end of August in the sidings near to Bedford St Johns and also on the Branston Junction triangle near Burton-on-Trent. On the 24th and 25th August the dynamic running bogie frame and underframe stress tests continued on the main line at speeds up to 90 mile/h hauling Test Car 6 and two British Rail Research Laboratory coaches number 6 (RDB975422) and number 10 (RDB975428) between Derby and Cricklewood via Manton Junction. On the 26th August a further test was operated at speeds up to 60 mile/h hauling a rake of 46 loaded HDA wagons totalling 2232 tonnes trailing load between Toton and Cricklewood, again via Manton Junction. On the 14th September static vertical loading tests were conducted in the EDU workshop whereby the locomotive body was raised off its bogies to just a sufficient height to have no load on the suspension, which was in order to establish the datum level for bogie strains. The dynamic bogie frame stress tests continued between the 19th and 20th September, again operating over the Midland Main line route to Bedford and Cricklewood; this time with the Class 58 plus Test Car 6 coupled with Test Car 10 and British Rail Research Laboratory Coach number 12 (RDB975136). The following day on the 21st September, a rake of 40 loaded HDA wagons was coupled with the Class 58 plus Test Car 6 and operated between Toton and Cricklewood. The final series of tests on the Class 58 that I was involved in, was the locomotive static and dynamic braking tests. On the 26th and 27th September following uncoupling of the Class 58 from Test Car 6, the locomotive ran light engine from Derby to Crewe to undertake light engine braking tests on the Crewe to Winsford level track section of the West Coast Main line.

Fig.194 - 59004 at the RTC Derby EDU workshop, January 1986

A snow covered Railway Technical Centre sidings at the end of January 1986 saw the arrival of the new Class 59 locomotives. Following shipment, the four locomotives (59001, 2, 3 and 4), built by General Motors Electro Motive Division (EMD), were hauled from Southampton Western Docks via Westbury to the Foster Yeoman depot at Merehead on 24th January 1986 for fuelling and commissioning. The following day all four new locomotives were hauled to Derby RTC by 47124. The locomotives were individually weighed on the RTC weighbridge, 59002 and 59004 were then hauled back to Merehead by Class 25 locomotive 25199 the following week. 59001 and 59003 remained in the midlands to undergo further acceptance testing, 59001 being taken to Toton diesel depot for load bank tests and 59003 was coupled with Test Car 10 (ADB975814)at Derby to provide a base for the test engineers and data recording instrumentation during dynamic testing along the Midland Main Line during early February 1986. Although I was not directly involved with the Class 59 acceptance tests, having worked on the design and development for the project during my apprenticeship there will always be something special about the Yeoman Class 59s for me.

Fig.195 - 59001 shortly after arrival at the Railway Technical Centre, January 1986

In the same manner as for passenger vehicles, body sway testing of locomotives was undertaken to verify what is called the Kinematic Envelope (KE). I was part of the team that carried out the sway test on the new Class 92 locomotive 92002 at the EDU workshop in Derby in December 1993. This test was typical of those undertaken in the 1980s and 1990s providing the design engineers with actual suspension movements from simulated effects of movement by tilting the locomotive and measuring the suspension movements and body-sway angles to support KE verification. In addition the suspension measurements an assessment of the pantograph sway was also evaluated during these tests. A sway test was carried out by jacking up one side of the vehicle under test in increments, securing each wheel onto packings at each stage, and taking measurements using a surveyor's theodolite to measure the exact position of the wheelsets, bogie frame and body at each lift increment. The method of measurements and analysis was developed by the DM&EE department in the mid-1980s and used prism reflector targets to accurately position the theodolite to take each measurement. Specially manufactured brackets were used to mount the reflector targets on both sides of the leading end axleboxes and bogie frame, extending laterally outside the vehicle gauge line to ensure they remained visible throughout the test. Further targets were mounted to the vehicle leading end at footstep level, cantrail level and also at defined measurement points on the underframe, normally at the drawgear drag box mounting bolt centres.

Fig.196 - Class 92, 92002 sway test, December 1993 [Servo]

A typical list of prism reflector targets is shown below in Table 17 which also included various reference targets used to evaluate whether the vehicle was swaying parallel to the rails and to ensure that the theodolite had not moved during the test.

Sway Test Targets	
1	Axlebox LHS
2	Bogie Frame LHS
3	Lower Vehicle Body LHS
4	Upper Vehicle Body / Cantrail LHS
5	Upper Vehicle Body / Cantrail RHS
6	Lower Vehicle Body RHS
7	Bogie Frame RHS
8	Axlebox RHS
9	Drag box Target LHS
10	Drag box Target RHS
11	Fixed Target (reference) Rail Level LHS
12	Fixed Target (reference) Rail Level RHS
13	Fixed Target (reference) EDU Workshop stanchion

Table 17 - Sway test target positions example

The theodolite was located some 20 metres away from the vehicle on a tripod, central to the track centre line. At each lift level a full set of measurements were taken by pointing the theodolite at each target in turn and recording the vertical angle, lateral angle and slope distance from the theodolite. In the early days these measurements were taken down by hand with the theodolite operator reading out the figures to a colleague who would write them down, then read back to verify as they were input manually into an HP basic computer. Later with the advances in computer technology, the readings were automatically taken via a direct wired RS232 type communications link between the theodolite and the computer, significantly improving both the accuracy and speed of processing the data. Safety during any testing was absolutely paramount therefore risk assessments were undertaken and detailed method statements written and briefed to all staff involved before the start of each test. Sway testing provided greater than usual challenges in developing and setting up safe systems of work to be adhered to, especially when jacking a 126 tonne locomotive. The safety equipment that was fitted between the running rails, in line with each wheelset was specially developed by the DM&EE test team to fit the EDU workshop tracks and prevented the risk of lateral wheelset movement whilst jacking. For a six axle locomotive such a Class 92 at least nine engineers were required to operate the test; an Officer in Charge, six persons, each operating a jack and installing the packings to raise the locomotive, one to operate the theodolite measuring equipment, and someone measuring and recording suspension clearances. It was critical during the test to maintain an equal lift across all axles whilst jacking so as not to cause unequal levels of sway at each end of the locomotive. The tests were also very time-consuming with a full jack to 12 inches of lift, in 2 inch steps on both left and right sides taking anything up to 12 hours, although this was reduced to about 8 hours when the direct measurements by computer were realised. The complex analysis computation process was developed by the DM&EE test team and fine-tuned over the years as computing technology advanced providing a quick output after each test for roll angles of the axle and bogie frame, and also the sway of the vehicle body at defined footstep, waist, cantrail and pantograph levels.

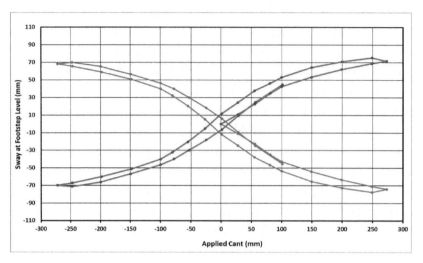

Fig.197 - Example sway test analysis

This sway test data analysis information was then used to verify the Kinematic Envelope for the vehicle under test; in the case of passenger stock this included verification from tests conducted with the air suspension in the normally inflated condition and also in the deflated state. For electric locomotives and electric multiple unit vehicles fitted with a pantograph the sway at the pantograph working height range was also confirmed.

Now and again Test Car 1 (TC1) was called upon for more high profile tests, such as locomotive acceptance tests of the new General Motors Class 59/2 locomotive 59201, built by General Motors Electro Motive Division (EMD) for National Power that arrived for test at the RTC Derby in March 1994. Prior to embarking on dynamic test running, static tests were carried out between the 7th and 10th March including wheel weigh distribution measurements, torsional stiffness ($\Delta Q/Q$), bogie rotational resistance, static brake characteristics and a parking brake test. Test Car 1 was used as the base for the test engineers and equipment for the first dynamic run with 59201 under its own power on UK tracks. Test Car 5, Test Car 6 and Test Car 2 were also included in the test train formation to provide additional brake power. The tests conducted on the first run out from Derby on the 14th March 1994 consisted of braking and riding performance, incrementing the speed up from 40 mile/h in 5 mile/h steps ensuring that the performance was within the required limits at each stage before accelerating faster up to the maximum designed operating speed of 75 mile/h. A check of the running stability was also carried out over short sections of the test running up to 80 and 85 mile/h.

Fig.198 - 59201 at Cricklewood, March 1994

On the morning of the test, the train formation with 59201 plus the four test coaches was hauled by a Class 47 from Derby to Nottingham so that the locomotive was facing south for the test run between Nottingham, Melton Mowbray, Manton, Corby and Kettering to Cricklewood. However because the locomotive was wired with instrument cables to TC1 the whole train had to be propelled between Dudding HIll Junction and Cricklewood Junction to turn the train before departure from Cricklewood back to Derby.

On return to the RTC the locomotive to test coaches and instrument cabling were disconnected in readiness for the tests to be conducted the following day with the locomotive running on its own for further brake and acceleration performance tests between Derby and Leicester. The route between Derby and Leicester was reasonably well suited to undertaking light-locomotive braking tests because there were a few sections of level track or with shallow gradients where a brake stopping distance test could be conducted to obtain valid results. The slow lines were used for tests at lower speeds up to 50 mile/h and the fast lines for the higher speeds. The braking tests undertaken with 59201 on the 15th March 1994 started off very well and luckily we had completed the majority of the tests by the early afternoon before the heavens opened; the last return trip from Leicester being extremely wet.

Fig.199 - 59201 at Leicester, March 1994

Soon after I joined the DM&EE testing section in 1985 a request was made by the design engineers to assist with investigations into reports of rough riding on Class 31/4 Electric Train Heat (ETH) conversion locomotives. Tests were carried out on two locomotives one of which was fitted with the modified suspension components to provide greater secondary suspension vertical bumpstop clearances.

Fig.200 - A Class 31/4 locomotive at Kettering, 1988

The tests were quite straightforward, and involved the use of a portable battery powered tri-axis vibration measuring unit that incorporated a thermal strip chart recorder. The unit was placed on the cab floor of locomotive 31467 and provided an instant paper trace record of the vertical, lateral and longitudinal acceleration levels measured during each of the various test runs between Doncaster, Peterborough and Lincoln. Additional test runs were completed for comparison using locomotive 31419 between Doncaster and Scarborough. The results of the tests were reported back to the design engineers for further assessment and following more modifications including trials with taperleaf type secondary suspension springs, various tests were carried out during late 1986 and again in September 1988 with locomotives 31409 and 31402 respectively; although I was not involved with these later tests I understand that the final modified suspension spring arrangement did somewhat improve the vertical ride of the ETH fitted Class 31 locomotives.

The Class 37 type locomotives had already been in service since the early 1960s however refurbishment along with various modifications were applied in during the 1980s. The refurbishment programme started in 1984 with the Class 37/4 locomotive conversions with Electric Train Heating (ETH) supply, followed in 1986 by the Class 37/7 type which had approximately 15 tonnes of additional weight added to provide better traction performance. An example of the modified Class 37/7 locomotives number 37799 was presented for brake testing in August 1986, and underwent static tests at the RTC before we undertook a dynamic braking test, running light engine from Derby to Manton Junction and back to confirm the acceptability of the brake stopping distance performance with the additional weight added.

More investigation work was carried out on Class 59 locomotive 59004 following reports of poor riding. On the 12th May 1994 a ride assessment test was conducted between Birmingham and Derby using portable ride meter located on the cab floor. Although I was not on board the test, I did undertake a bogie rotational resistance test on both bogies of 59004 on arrival at the RTC. As a result of the findings of the bogie rotation test, the Foster Yeoman maintenance team, assisted by engineers from the Engineering Development Unit (EDU), lifted the locomotive off its bogies at the Derby workshop on the 24th May. The lining material of the centre pivot was then replaced and a further bogie rotation test was then completed to confirm the expected improvements in rotational torque had been realised. I undertook the subsequent dynamic ride assessment between Derby and Water Orton on the 3rd June to measure the level of improvement in the riding properties using the portable ride meter. On reaching Water Orton I alighted and 59004 continued on its way back to Merehead.

During 1995 I was again involved in undertaking ride assessment work on Class 59 locomotives, this time in regard to the evaluation of the effectiveness of additional bogie yaw dampers that were installed as a trial on locomotive 59103 on 20th September. Again the portable ride assessment equipment was used and test runs were conducted over a three day period based at Merehead. The first day's testing involved 59004 in standard condition to record data as a baseline; we completed two round trips between Merehead, Westbury and Salisbury. Day two was with modified 59103 with two return runs on the same route, and the third day with both locomotives shuttling between Merehead and Westbury.

As part of a project to assess and compare the acceptable levels of vibration dose levels that traincrew were subjected to during their working day, a series of tests was undertaken on many different types of rolling stock.

Fig.201 - Class 73 locomotive 73210 during test, April 1988

One such test I conducted was a test run undertaken on the 12th April 1988 using the portable ride acceleration measuring equipment; on this occasion a Class 73 locomotive was the subject undergoing test on two return trips between Stewarts Lane depot and Orpington coupled with a Class 423, 4-VEP Electric Multiple Unit to provide additional braking power whilst testing up to the maximum operating speed of 90 mile/h. An example of the data analysis output from a similar vibration dose test, in this case a vertical floor level measurement is shown in Fig.202 below showing a time history trace of the 100 Hz low pass filtered floor level acceleration and the cumulative dose exposure level.

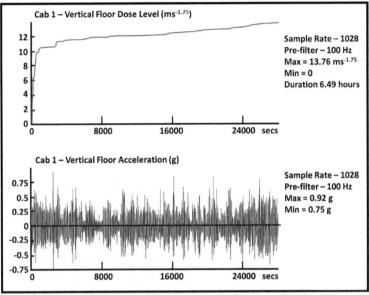

Fig.202 - Example vibration dose level analysis [Serco]

The analysis process involved filtering the acceleration signals using a British Standard defined weighting characteristic relating to whole body vibration assessment, then cumulating with respect to the time period. The maximum dose for the example 6.49 hour test duration was 13.76 ms$^{-1.75}$ which, when combined with the lateral and longitudinal vibration dose levels, and factored up to estimate the dose over an 8 hour shift gave a value of 14.5 ms$^{-1.75}$. Quite clearly in this example the vertical acceleration levels predominate. The limiting criteria in place at the time mandated that traincrew must not be subjected to a combined vibration dose level greater than 15 ms$^{-1.75}$ during a working day. The maximum allowable working hours before reaching the dose limit was then derived from this to be 9.2 hours.

Cab Floor Vertical Acceleration - Vibration Dose	
Maximum Train Speed	52.6 mile/h
Test Duration	6.48 hours
Maximum Vertical Vibration Dose	13.76 ms$^{-1.75}$
Maximum Later Vibration Dose	1.46 ms$^{-1.75}$
Maximum Longitudinal Vibration Dose	0.26 ms$^{-1.75}$
Overall Vertical / Lateral / Longitudinal Dose	13.77 ms$^{-1.75}$
Dose value for working 8 hour day	14.5 ms$^{-1.75}$
Maximum allowable duration	9.2 hours

Table 18 - Example vibration dose combined assessment

The locomotive tested (73210) was retired from main line use and preserved in 2008, and at the time of writing was residing at the North Norfolk Railway.

Occasionally there were rather more prestigious visitors to the RTC Derby, such as steam locomotive Flying Scotsman (4472) which arrived under its own power in the early hours of a misty 12th December 1986 for wheel weighing using the EDU weighbridge. Such a visitor generated considerable interest from staff working at the RTC who took a break from work in the many offices on-site to come down the workshop to take a look and soak up a bit of the nostalgic atmosphere brought by a steam locomotive on site.

Fig.203 - Flying Scotsman at the Derby RTC, December 1986

MORE UNUSUAL VEHICLES AND TESTS

SPECIAL PURPOSE VEHICLES

During the 1980s there were a number of new designs of on-track plant vehicles developed, many being self-powered vehicles incorporating up to date technology to support the maintenance and renewal of the infrastructure. Where the vehicles had a reasonable self-powered designed operating speed (for example 60 mile/h), the testing of such vehicles normally involved combining of ride, braking, traction performance and environmental tests during a self-powered test run between Derby and Manton or Leicester. Whereas some vehicles only had the ability to move under their own power at slow speeds, meaning that the dynamic testing of such vehicles had to be carried out in the same manner as a freight vehicle in a test train hauled by a locomotive.

Fig.204 - Plasser SCPV14 piling machine at Bedford sidings, July 1986

In July 1986 a strange looking new plant vehicle (LDRP96511) arrived in Derby for acceptance testing. The SCPV14 (Self Contained Piling Vehicle) built by Plasser & Theurer was self-propelled for slow speed movement only and designed specifically for the installation of piled foundations for the East Coast Main Line (ECML) electrification. An on-board crane was used to lift a hydraulically operated pile hammer with a pile attached, then position alongside the line where the electrification mast was required, and hammer the pile into the ground.

Apparently it was a very noisy affair, but reports from its operation were that pile installation work at speeds up to 5 piles per hour could be achieved. The design speed for haulage by a locomotive was 60 mile/h, however a speed of only 5 mile/h was possible when self-propelled, which was sufficient to move between piling positions within a work-site. The purpose of the tests undertaken was to confirm the static test compliance followed by running characteristics (ride), braking and traction performance. The static tests at the EDU workshop confirmed that the torsional stiffness $(\Delta Q/Q)$ properties were good, but the bogie rotational resistance 'X' factor levels were slightly lower than expected. A judgement was taken to continue with the dynamic testing, with Test Car 1 being used for the tare condition ride test between Derby and Bedford on the 14th July 1986, hauled unusually by a Class 31 number 31418; unfortunately the lateral riding properties of the SCPV14 machine at speeds above 50 mile/h was found to be somewhat unstable, with the 3-piece bogies having a tendency to hunt, which was probably not helped by the slightly low bogie rotational resistance values. A laden condition ride test was completed successfully on the following week between Derby and Bedford, followed by traction and braking performance tests which were conducted using Test Car 2. For the traction tests the train was hauled to the Egginton Junction to Mickleover, British Rail Research operated test line; this 8 mile long line was well suited to such tests due to the 1½ mile long section of 1-in-110 rising gradient. In order to measure the traction performance a strain gauged coupling was used that had been calibrated as a load cell; this was installed between Test Car 2 and the SCPV14 machine.

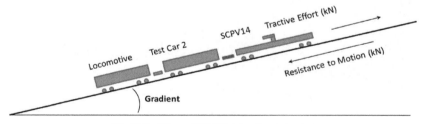

Fig.205 - Plasser SCPV14 traction test set-up

The Test Car and the Class 47 test train locomotive, number 47049, were used as the trailing load for the SCPV14 to haul during the tests, totalling approximately 154 tonnes. The coupling loads measured during the tests were recorded onto a UV paper chart recorder, and the values measured at varying speed increments extracted into a table of results. An example of a performance characteristic plotted from the equivalent drawbar tractive effort results is shown in Fig.206.

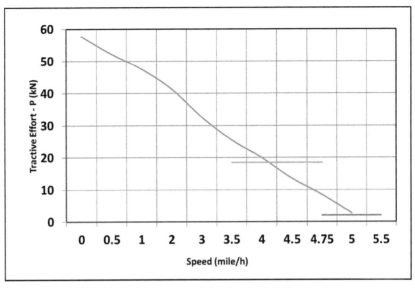

Fig.206 - Plasser SCPV14 tractive effort

After completion of the Mickleover tests the train was returned to the RTC Derby and the strain gauged coupling was removed. Further braking tests for the SCPV14 were undertaken between Crewe & Winsford using the normal slip/brake method for assessing the brake stopping distance performance up to 60 mile/h with Class 47 locomotive 47125 in charge. Due to the lateral riding performance, the speed at which the SCPV14 Piling Machine was initially accepted for operation was restricted to 45 mile/h when in the tare condition. The SCPV14 did however return later in 1988 following modifications, for a repeat test that carried out between Derby and Bedford on the 29th September, which demonstrated acceptable performance to allow operation in trains at speeds up to 60 mile/h in both the tare and laden conditions.

In mid-1995 the Engineering Development Unit (EDU) was entrusted with the conversion of the first of the new Generator Van vehicles designed for use with the Nightstar sleeper vehicles. The requirement emerged from the need to supply power to the new Nightstar sleeper vehicles when they were being operated north of London hauled by Class 37/6 locomotives, because the Class 37 locomotives could not provide the 2500 volt electric train supply required for the Nightstar vehicles.

Fig.207 - The first of the generator vehicles 6371 + 37602, July 1995

The Generator Van vehicles were designed around redundant Mk III (SLEP type) sleeper coaches; the first conversion used coach number 10545 which was subsequently renumbered to 6371. Modifications started by stripping out the original interiors and strengthening the vehicle body structure to allow the installation of two high powered diesel generating sets. In order to install the generator sets large apertures had to be cut into the roof at both ends of the Mk III bodyshell. Fuel tanks were installed and the suspension was up-rated to cater for the extra weight. Blue-star multiple working cables and locomotive control air connections were also provided through the Generator Van, this allowed one of these Generator Van vehicles to be sandwiched between two of the newly converted Class 37/6 locomotives operating in multiple.

Fig.208 - 37601 during preparations for testing at the RTC Derby, July 1995

Once the EDU team had completed the modifications, the Generator Van was handed over to the testing section to commence acceptance tests, starting with wheel weighing, static bogie rotation, torsional stiffness (ΔQ/Q) tests and body sway tests; all of which were carried out with the air suspension in both the inflated and deflated conditions. Static brake tests were also carried out including confirmation of the brake application and release timings and brake pad force measurements. The dynamic test runs commenced on the 30th July 1995 with Generator Van 6371 undergoing slip/brake testing between Crewe and Winsford using Test Car 2. Two of the newly converted Class 37/6 locomotives were then supplied to the RTC Derby to support the remainder of the testing up to a maximum speed of 80 mile/h, coupled to either end of 6371 in a representative manner to which the formation was due to be operated in service. Test instrumentation was fitted to 6371 to independently measure speed and distance, accelerometers and suspension displacements were installed for running characteristics assessment and pressure transducers fitted to monitor the brake cylinder pressures and air suspension pressure during dynamic running. The data recording equipment being installed into the No.2 end cab of 37601 which provided the base for the test engineers during the dynamic running tests. Further brake stopping distance tests were carried out on the 9th August 1995, this time with the Generator Van coupled within the formation 37602 + 6371 + 37601, operating three round trips between Derby and Leicester. On each run we undertook both full service and emergency brake stops at pre-determined locations on the route where the gradient of the line was shallow enough not to affect the assessment of the brake performance of the formation. The reason for undertaking braking tests coupled between the two Class 37s in addition to the slip/brake test with the Generator Van vehicle on its own, was to evaluate the performance of the short formation Class 37 + Generator Van + Class 37, and how the brake balance between the cast iron block braked locomotives worked with the disc braked Generator Van. The test results showed that whilst the overall braking performance of the formation was acceptable, the braking characteristics of the disc braked Generator Van provided greater levels of brake effort at higher speeds compared to the cast iron block brakes on the locomotives. This undoubtedly affected the life of the brake pads, therefore it was recommended that the brake pad wear rates should be monitored once the Generator Van entered service. The final series of tests from the 16th August commenced with the formation 37601 + 6371 + 37602 heading north from Derby undertaking riding characteristics measurements between Derby and Doncaster. The riding measurements were carried out by using accelerometer transducers fitted to the underframe inside the generator coach, both vertical and lateral plane measurements being recorded above each bogie centre and also in the centre of the coach. The following day being coupled to the Doncaster Works test train formed of 7 Mk II coaches, and operating further running characteristics test runs between Doncaster and York completing three round trips, with the Class 37 + Generator Van + Class 37 formation running round the coaching stock for each return run.

ROAD AND RAIL FREIGHT WAGONS

In early May 1986 a new design of freight wagon arrived at the Railway Technical Centre by road. The vehicles were lorry trailers that were built-up by coupling the trailers in between bogie units (TOPS code PXA) to form an articulated freight train. The vehicle was designated Trailer-Train. A total of five trailers were produced, three of which were provided to the RTC for testing along with four GPS type bogies and two adaptor units that were fitted with standard buffers and drawgear. The Trailer-Train was designed for operation at speeds up to 75 mile/h in the tare condition and 60 mile/h when fully laden.

Vehicle	Production	Test Train Formation
PXA 'Trailer Train' Bogie	PXA 'Trailer Train' Bogie	TN 95901 - 95904
PXA 'Trailer Train' Adaptor	PXA 'Trailer Train' Adaptor	TN 95951 - 95952
PXA 'Trailer Train' Van Trailer	PXA 'Trailer Train' Van Trailer	TN 96001 - 96003

Table 19 - Trailer Train Vehicles

The Trailer Train arrived at the RTC with the bogies in pairs on two low loader lorry trailers, followed by the three Trailer Train lorry trailers, therefore the hard standing area outside the front of the EDU workshop was busy for a few hours getting the bogies onto the track and the train assembled ready to start testing.

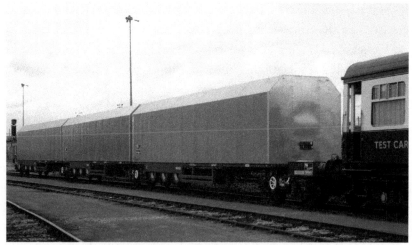

Fig.209 - Trailer Train at the RTC sidings, May 1986

The assembly process started with lowering the rear bogie (furthest from the test coach) onto the track with its buffer adaptor on top, followed by reversing the rear of the first lorry trailer into position with the adjustable suspension raised to its highest setting, and locking into the bogie adaptor. The lorry tractor was removed, the road wheels suspension retracted and with the front of the trailer supported on its landing legs, the second bogie rolled into place, locking into the trailer 5th wheel. The process of attaching the next two trailers followed in the same manner, before locking into place the front bogie fitted with the buffer adaptor unit, completing the 3-vehicle unit that was over 42 metres long. Connection of the air pipes between each adaptor, bogie and trailer was completed, a static brake continuity test carried out using the Class 08 shunt locomotive, and the train shunted round to the weighbridge road at the EDU for wheel weighing. Following completion of the requisite torsional stiffness ($\Delta Q/Q$) and bogie rotation tests the Trailer-Train was instrumented for ride testing, coupled with Test Car 1, and a dynamic test undertaken in the tare condition between Derby and Cricklewood on the 8th May 1986 hauled by 47063; a maximum speed of 85 mile/h was attained during this first test run. Static brake tests were then carried out including brake pad force measurements on the single disc per axle arrangement fitted to the Trailer Train. A tare condition slip/brake test was carried out between Crewe and Winsford on the 15th May up to the designed maximum operating speed of 75 mile/h. In order to facilitate the loading process before the laden tests, the whole train had to be disconnected, and each lorry trailer was taken away from the RTC site by road to be filled with pallets of a well-known brand of dog food. Once returned to the RTC and re-formed, all air pipes connected and a brake continuity test carried out, the laden the Trailer-Train was re-coupled and attached to Test Car 1 for another ride test on the 16th June. This test run was carried out between Derby and Bedford being hauled by locomotive number 45122. The use of a Class 45 locomotive on test train by 1986 was unusual, as the fleet of these locomotives was in declining numbers, 45122 being withdrawn from service less than a year after use on this test. The normal traction for the ride testing runs was by now the Class 47 locomotive type. For a final test run, the Trailer-Train was coupled Test Car 2 for a laden condition slip/brake test on the 19th June 1986.

A second generation of the Trailer-Train type vehicles was produced in 1991, and being slightly different in tare weight, the modified design required re-testing in that condition. A three vehicle train was submitted for test to confirm torsional stiffness, bogie rotation and riding properties. The testing was undertaken during

July 1991, successfully completing a dynamic ride test run between Derby and Cricklewood on the 18th July.

In mid-1991 another new road trailer type freight train, designed by Cravens / Tasker for Charterail, was submitted for test. The concept was very similar to the Trailer-Train design seen a few years earlier; however this design used Y25 type bogies fitted with adaptor units, to which the lorry trailers were coupled into forming the train. Three trailers along with four bogies and the associated adaptor units were delivered to the RTC Derby in early June.

Fig.210 - Charterail road rail train - RTC Derby, June 1991

After coupling together and completion of the static tests, an initial ride test carried out on the 24th June in the tare condition, between Derby and Kettering, highlighted an issue with contamination of the suspension friction damper surfaces. Following completion of the test and removal of the test instrumentation cables, the train was disconnected and the bogies dismantled at the EDU workshop to degrease the dampers. Whilst this rectification work was being undertaken the three lorry trailers were sent away by road for loading, again with pallets of dog food. The loaded lorry trailers were returned to the RTC, re-formed and coupled to Test Car 1; we then re-attached all the instrument cables and re-checked everything was working in the test coach.

Vehicle	Production	Test Train Formation
KDA Bogies	CRL 96101-96104	CRL 96101-96104
KDA Adaptor (buffers)	CRL 96201	CRL 96201
KDA Adaptor (intermediate)	CRL 96221-96222	CRL 96221-96222
KDA Adaptor (buffers)	CRL 96281	CRL 96281
KDA Rail/Road Trailer (1991) end load type	CRL 96301-96303	CRL 96301-96303
KDA Rail/Road Trailer (1992) curtain side type	CRL 96304-96306	CRL 96304-96306

Table 20 - Charterail Vehicles

A second ride test commenced on the 1st July 1991, running from Derby along the normal route towards Manton Junction. All was going well, and the laden riding properties of the Charterail train were much improved from that of the initial tare test; however passing Melton Mowbray at about 10 am, the signaller stopped the train just to the south of the station, reporting seeing sparks emanating from the rear wagon, and that it appeared the curtain sides of the wagon had come open.

Fig.211 - Charterail train following failure at Melton Mowbray, July 1991 [Serco]

We ensured the line was safe with all other trains were stopped by the signaller, and walked back along the cess side of the line to discover that the rear lorry trailer structure had collapsed; the trailer was being supported by its landing legs which were resting on top of the rails. Luckily the rear bogie was undamaged, and in order to clear the line whole train was moved at walking pace, whilst under constant inspection, and shunted into the up goods loop at Melton Mowbray, where a more detailed inspection was carried out. The test run was abandoned there because the vehicle was deemed unsafe to move any further; subsequently we disconnected the test instrumentation and uncoupled Test Car 1. The test train locomotive then returned the Test Car and ZXA van to the RTC. The detailed inspection that followed confirmed the failure of the lorry trailer structure was along the rivet jointed side panels of the trailer. The vehicles were subsequently unloaded and disconnected on-site before being recovered by road back to the manufacturer.

Following a complete re-design of the trailer structure to a curtain sided type, three new lorry trailers were re-submitted about a year later for a retest, and formed together coupled with the original set of bogies and adaptors. Strain gauges were applied to the lorry trailer structure before the trailers were loaded with steel and concrete block test weights. A ride test run was then carried out between Derby and Bedford on the 6th August 1992, successfully operating up to 75 mile/h. The following week after measuring the static braking characteristics, were took the loaded three vehicle train to Crewe for slip/brake testing, again successfully completing the tests up to 75 mile/h.

The train was released for use in service in the loaded condition only, but returned again later in the year for further detailed investigations into the structural integrity of the lorry trailers following a period in service. During these tests in October and November 1992 we also undertook repeat tare condition testing with the new design of lorry trailers. Static torsional stiffness ($\Delta Q/Q$), bogie rotation tests and installation of over 30 strain gauges onto the lorry trailer structures was completed between the 26th and 30th October 1992. The train of three vehicles was then loaded again with steel and concrete block test weights and a ride / stress test conducted on the 3rd November between Derby and Bedford. On return to Derby the trailers were unloaded, and a tare condition ride test successfully completed on the 5th November between Derby and Cricklewood. Data analysis from the stress measurements during the loaded test run prompted the design engineers to request further static loading tests to record the stress levels in the lorry trailer underframe structure. These static tests were carried out with varying levels of load applied at the EDU workshop between the 24th and 26th November. It is believed that the vehicles only operated for a short period in service after the final tests were completed.

RESISTANCE TESTING

In October 1986, I supported the DM&EE traction test team undertaking train resistance measurement tests on the Western region. These tests were carried out between Bristol and Taunton using Test Car 6, a rake of HAA wagons, and a standard 20 Ton brake van hauled by a Class 47 locomotive number 47284.

Fig.212 - 47284 with Test Car 6 at Taunton, October 1986

The remit for the tests was to accelerate the train from Bristol heading towards Taunton, and at 60 mile/h when passing a predetermined location near Yatton, the power was shut off allowing the train to coast until it came to a stand. The test section between Bristol and Taunton was chosen because the gradient profile of the line was essentially level for a 20 mile section between Yatton and Bridgewater; the lack of gradients making it much easier to calculate the rolling resistance of the train.

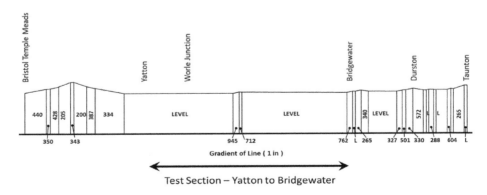

Fig.213 - Route gradient profile representation, Bristol to Taunton

Before setting off from Bristol the traction inspector contacted the signaller to ensure the whole route section was cleared with green signals such that the complete run could be attained without the need to apply the brakes. By allowing the train to coast naturally to a stand without using any braking, an assessment of the train rolling resistance was established by computing the distance and time records taken from each run.

Fig.214 - Test Car 6 at Taunton, October 1986

Luckily the weather conditions were kind to us on the two test days being dry and with very little wind, which also helped with the evaluation of the results. On each of the two test days, only two test runs were carried out due to the length of time each test took to complete. Yes it does take a long time for a train to come to rest if you do not use the brakes.

BRITISH RAIL RESEARCH VEHICLES

The British Rail research department had many vehicles that were purpose built for supporting the development of engineering solutions such as new braking systems and suspension systems. An example of such a vehicle was RDB511023 built in the mid-1960s.

Fig.215 - HSFV1, RDB511023 at RTC, July 1987

This was a two axle vehicle designated HSFV1 (High Speed Freight Vehicle 1), that was used over many years for the development of new suspension systems including that which was eventually incorporated into the two axle Railbus vehicles in the late 1970s. Further modifications were made to the vehicle in the mid-1980s, and although the DM&EE test section was not directly involved in the suspension development, in 1985, following the fitting of flexicoil type suspension springs, I lead a DM&EE operated ride test run with the vehicle on the midland Main line. The vehicle, loaded with packs of rail was coupled to Test Car 1, and on the 17th September 1985 a test run was carried out between Derby and Cricklewood at speeds up to 100 mile/h. The Class 47 locomotive used for the test run had received a special maintenance safety exam prior to being authorised to operate in this test up to 100 mile/h as opposed to the normal maximum speed of 95 mile/h for Class 47 locomotives.

In mid-1985 I undertook a ride assessment on the Structure Gauging Train (SGT) optical measurement vehicle using portable test equipment. This unusual looking vehicle was a specially designed and built vehicle housing the structure gauging optical measurement system developed by the British Rail Research department. The vehicle was permanently coupled between the Laboratory support coaches forming what was known as the SGT, comprising of three vehicles:-

DB975081   a Driving Trailer converted from Laboratory Coach 17 - Hermes

DC460000   the Optical Measurement Car sometimes referred to as the elephant car

DB975280   a Support Coach, converted from Laboratory Coach 18 - Mercury with accommodation and staff facilities.

The driving trailer vehicle allowed the operation with the test train locomotive remaining attached to the opposite end of the train to work in push - pull mode; the locomotive being controlled via the standard blue star multiple working 27-way jumper cable system. The test runs that operated out and back from Derby along the Midland Main Line were successful, and the train entered service operated by the Director of Civil Engineering (D of CE) department gathering structure information across the network.

Fig.216 - Structure Gauging Train at the RTC, July 1987

The last of the Class 150/1 DMU series units to be built at York in 1987 was specially adapted as a Track Recording Unit (TRU) comprising of vehicles DB999600 and DB999601. The two car unit was fitted out with the recording equipment at RTC Derby in 1987 and submitted to the DM&EE for static testing to confirm the wheel weight distribution, torsional stiffness and bogie rotation compliance before entering service with the Director of Civil Engineering (D of CE) department.

Fig.217 - Class 150/1 based Track Recording Unit at the RTC, 1987

Both these trains are, at the time of writing still in operation as part of the Network Rail infrastructure recording train fleet.

CHANNEL TUNNEL TESTS

In September 1994 the test team were called upon to undertake tests on the new Euro Tunnel le Shuttle trains.

Fig.218 - The author on a Euro Tunnel le Shuttle, September 1994

After preparing a set of instrumentation and long lengths of cabling in Derby, we travelled by van via the ferry from Dover across to France and then on to the Euro Tunnel F40 maintenance depot in Coquelles. A Euro Tunnel le Shuttle train formed of both Single Deck (SD) and Double Deck (DD) road vehicle transporter wagons was provided specifically for the tests. The wagons were loaded with a variety of road vehicles before the instrumentation and cables were installed. The SD wagons had a mixture of cars and articulated lorries on-board whilst the DD wagons had only cars loaded.

le Shuttle Vehicle Type	ID / Instrumented	Vehicle Number
Locomotive	1	9027
Single Deck Loader	2	3811
Single Deck Car Carrier	3	3003
Single Deck Car Carrier	4	3410
Single Deck Car Carrier	5	3206
Single Deck Car Carrier	6	3006
Single Deck Car Carrier	7	3711
Single Deck Car Carrier	8	3007
Single Deck Car Carrier	9	3008
Single Deck Car Carrier	10	3706
Single Deck Car Carrier	11, Instrumented	3210
Single Deck Car Carrier	12	3202
Single Deck Car Carrier	13, Instrumented	3703
Single Deck Car Carrier	14	3011
Single Deck Loader	15, Instrumented	3810
Double Deck Loader	16, Instrumented & test base	1811
Double Deck Car Carrier	17	1209
Double Deck Car Carrier	18, Instrumented	1420
Double Deck Car Carrier	19	1012
Double Deck Car Carrier	20	1213
Double Deck Car Carrier	21	1402
Double Deck Car Carrier	22	1017
Double Deck Car Carrier	23	1210
Double Deck Car Carrier	24	1715
Double Deck Car Carrier	25	1002
Double Deck Car Carrier	26	1216
Double Deck Car Carrier	27	1714
Double Deck Car Carrier	28	1008
Double Deck Loader	29	1818
Locomotive	30	9031

Table 21 - Channel Tunnel Shuttle test train formation

Fitting the test instrumentation involved a significant amount of walking up and down the length of the 28 wagon le Shuttle train formation; so not surprisingly I had ample opportunity to record all the wagon numbers which are listed in Table 21 below. The test base, signal conditioning and data recording equipment was installed in the crew area which was in the upper level at the end of the DD loader wagon number 1811. The instrumentation installed consisted of 21 accelerometers located across five wagons to record the bogie and vehicle body stability, ride performance as well as the comfort levels. The tests were conducted over a three day period between 23th and 25th September, during which we made a total of ten round trips through the tunnels between Coquelles and Folkestone at speeds up to 160 km/h (100 mile/h). Before the testing on the last day, the cars and articulated lorries were unloaded and the last test runs were completed with the wagons in the tare condition. One test run was carried out with a bogie yaw damper removed to assess the running stability levels in the unlikely event of a damper failure. Following successful completion of the tests on the 26th September we removed the instrumentation and cables before making our way back to the UK via the ferry.

Later in 1994 I was again involved in tests in the Channel Tunnel, this time with the Class 92 locomotives which had been in service on British Rail lines for about a year. The tests were primarily performance trials to establish suitability for hauling freight trains through the tunnel to and from France. These tests made use of Test Car 2 which was taking a well-earned break from the regular slip/brake testing, and still sporting its smart British Rail Maroon colours. The heavy duty draw gear fitted to Test Car 2 also being particularly suited to the coach being incorporated within heavy freight test trains. The instrumentation installed for these tests included pantograph uplift force measuring equipment to measure current collection performance, a strain gauged coupling to measure drawbar tractive effort and pressure measuring equipment to record the aerodynamic pressure pulses whilst operating through the tunnels. The pantograph uplift force measuring system incorporated telemetry equipment to transmit the data signals from the pantograph that was live at 25,000 volts, to the signal conditioning equipment in the test car. Roof mounted video cameras and lighting provided visual monitoring of the pantograph to overhead wire interface during operation; the VHS video recordings of these images had an overlay image incorporating the time and train speed which aided post-test review of the video footage.

Fig.219 - Test Car 2 at Dollands Moor, December 1994

A considerable amount of the instrumentation was installed onto locomotive 92003 and into Test Car 2, needing an additional tall instrument rack fitting into the saloon area which was carried out at Derby during mid November 1994. The Class 92 and Test Car 2 were hauled to Crewe on the 21st November by a Class 47 locomotive and preliminary tests were carried out on the 22nd November on the West Coast main line to measure pantograph current collection performance between Crewe and Carlisle before the formation of 92003 and Test Car 2 was hauled south to Dollands Moor freight yard where some final instrumentation setting up had to be completed over two night shifts on the 28th and 29th November. The small workshop building at the West end of the Dollands Moor sidings only being long enough to accommodate one vehicle did not provide much protection from the cold winter nights, therefore we made good use of Test Car 2s excellent storage and fan heaters, not to mention the baby belling cooker for a welcome bacon-butty snack.

Fig.220 - 92003 and 92018 at Dollands Moor, December 1994

On the 1st December a second Class 92 locomotive number 92018 was provided and coupled into the test train formation and some additional instruments prepared ready for the start of testing. The test runs commenced on the 12th December with a train of loaded container flat wagons totalling over 1500 tonnes of trailing load, operating between Dollands Moor and through the Channel Tunnel to the Fréthun freight yard near Calais. The test runs included constant speed running and also standing-start pull-away tests on the steepest part of the rising gradient inside the tunnel approaching the UK with both locomotives operating, in degraded power configurations with only one locomotive powered and also with only one bogie of one locomotive powered. The Class 92 tests based at Dollands Moor operating through the tunnels to France continued until the 15th December, following which the formation was hauled back to Derby for de-instrumentation and the return of Test Car 2 to its normal state to enable slip/brake testing to resume after the Christmas break.

INVESTIGATIONS AND MODIFICATION ASSESSMENT

In many cases the testing that the DM&EE test team were called upon to undertake was to investigate problems with older vehicles, such as Sturgeon Type 'A' wagons built in the mid-1950s by Head Wrightson Ltd of Thornaby. These wagons were fitted with diamond frame fabricated 3 piece bogies and plain white metal bearings; most of the wagons which were built for the engineers department did not have any train braking system fitted (i.e. they only had handbrakes).

Fig.221 - 31415 + Test Car 1 + Tench and Sturgeon wagons at Corby, April 1988

During the early to mid-1980s many of the Sturgeon wagons were converted for different uses, such as for carrying track components rather than general materials; the conversions which included the fitting of air brakes resulted in some vehicles being re-designated as Tench wagons. Both Sturgeon and Tench wagon types retained the original 50 tonne load capacity, however the change of use along with the increased operational flexibility after fitting with air brake, gave rise to increased reports of axle bearing failures. In March 1988 two of these type YBA civil engineers wagons, DB994441 which was a modified Sturgeon flatbed, and DB994267 with original dropside doors and metal ends designated as Tench, were presented at the RTC Derby for assessment of axle bearing temperatures. Two test runs were carried out, the first on the 29th March 1988 with the Tench wagon DB994267 in the tare (no payload) condition, and the Sturgeon wagon DB994441 in the fully loaded state. Before the second test run on the 6th April, the load condition of the two wagons was reversed, with the payload of concrete sleepers being transferred from the Sturgeon wagon onto the Tench wagon using the overhead cranes at the EDU workshop in Derby. Both the test runs were undertaken between Derby, Manton, Corby and Kettering, the locomotive running round the test train in the old station yard sidings at Kettering before retuning via Manton to Derby. The test speeds were limited due to the poor riding properties of both the wagons and the rise in axle bearing temperatures at speeds above 40 mile/h especially on the respective loaded wagon. The tests were therefore not very successful in supporting any increase in the 35 mile/h operating speed for both types of wagon.

Fig.222 - Test train at Kettering ready for the return trip to Derby, April 1988.

July 1985 started off with a trip to Merehead Quarry to assist the design engineers and Foster Yeoman with the evaluation damage to aggregate wagons drawgear, which was being seen on the relatively new 102 tonne aluminium bodied PHA hopper wagons built by Procor.

Fig.223 - 102 tonne wagon PHA 17827 at Theale, July 1985

The damage that was occurring to the instanter coupling links and the drawhook rubber stacks mounted behind the headstock, which was indicative of high shocks or loads within the drawgear at some point during the traffic flows where the new wagons were being used between Merehead Quarry and Theale. A series of tests were set up to establish possible causes of the drawgear damage, primarily monitoring coupling movements and longitudinal acceleration shocks at different points along the train whilst operating in the normal service runs between Merehead and Theale hauled by Class 56 locomotives. On the day of the tests on the 3rd July 1985 locomotive 56039 was provided from Westbury for the duty, and we joined the train at Merehead early on the very warm summer's morning to fit the accelerometer data recorder instruments located on the access step platform of three wagons at different positions in normal service train. The longitudinal acceleration data recorders were set running just before the train departed and riding in the Class 56 locomotive cab we kept a log of the speeds, power and brake demand and any instances where longitudinal shocks were felt. Upon arrival at Theale we returned to each data recorder and extracted the paper trace results. The recorders were re-started ready for the shunt movements into the Theale sidings for unloading of the wagons. Due to the overall sidings length the train had to be split into two parts to facilitate the unloading. When the train was propelled back into the sidings at slow speed, as the brakes were applied to bring the train to a stand when nearing the end of the siding, there was a large longitudinal shock towards the rear of the train. It was established that as the brakes were applied from the locomotive at the front of the train, due to the slow speed and the very slight rising gradient into the siding, the front of the train came to a stand before the brake had propagated to the rear of the train, therefore the rear few wagons stopped suddenly as the couplings tightened causing a shock.

Fig.224 - Instanter type couplers, short position (Left), long position (Right)

The shunting and propelling movement was repeated a number of times to confirm what was being seen, and in some cases it was noted that the shocks were so great to cause the instanter links a coupling near the rear of the train to move to the long position; when a repeat shunt movement was performed with a coupling in the long position then the shock was even greater. The effect of the shocks was seen in the damage to the drawhook rubber stacks that were mounted behind the headstock to help prevent coupling shocks from being transmitted through the wagon underframe. It can be seen from the photos in Fig.225 the difference between a rubber stack on wagon PR17827 in good condition and the one on wagon PR17834 that was damaged by large shocks of loads.

Fig.225 - Drawgear rubber stacks, good (Left), poor (Right)

Fig.226 - 56039 at Theale, July 1985

Following the series of tests undertaken in July 1985, further investigations were carried out during early 1986 into the ongoing damage issues to the drawgear on the 102 tonne aluminium bodied aggregate hopper wagons. These tests included an assessment of the potential benefits of AAR type knuckle couplers and also looked at a prototype device to apply the brake at the rear of the train when propelling into the sidings at Theale; the idea being that the brake propagating forwards along the train rather than backwards eliminated the shocks when propelling.

Fig.227 - Procor 102 tonne aggregate hopper wagon PHA17832 Theale, March 1986

The evaluation and comparisons between the coupling types were realised by using PTA type (ex Iron Ore tipplers) wagons fitted with AAR knuckle couplers, and the new PHA type 102 tonne aluminium wagons which were fitted with conventional buffers, drawgear and instanter couplings.

Fig.228 - Ex-Iron ore wagon PTA26820 with AAR couplers Westbury, March 1986

The tests involved the operations with various train formations from Westbury Yard to Theale using two Class 56 locomotives, numbers 56039 and 56048. Portable test instruments fitted with internal accelerometers were secured to the wagon headstocks at intervals along the train for measuring any longitudinal shocks that occurred during the test runs. Each test run was operated in a normal service train path along the 55 mile main line route from Westbury, over Savernacke summit, through Bedwyn and Newbury to Theale.

Fig.229 - 56039+56048 awaiting departure from Westbury, March 1986

The gradients on the route were quite challenging for the heavier freight trains, which was why two Class 56 locomotives were commonly used until the Class 59s were introduced. Three days of testing were carried out, firstly with the train formed of PHA wagons, then a day of testing at Theale investigating the effects of the brake device, then a third day with a train of PTA wagons. The results from monitoring the longitudinal shocks along each of the train types during the test runs from Westbury to Theale were compared. It was notable that the shocks and coupling snatching seen in the train of 102 tonne PHA wagons were greater and more frequent than those seen in the PTA wagons train. The shocks were more prevalent when pulling away from stationary; however on a couple of occasions instances were noted at speed, when applying power again after using the brake to check the speed. This was predominantly due to the slack in the instanter type couplings between each PHA wagon, whereas the AAR knuckle type couplers did not allow such longitudinal movement between wagons. The assessment of the shocks and coupling snatching were also assessed during the normal process of shunting the train at Theale, again it was noticeable how much smoother the train movements were with the PTA wagons train. The brake application device trialed on the second day was fitted to the rear of the last PHA wagon before the train was propelled into the siding; it consisted of a plastic tube with a cap end to which the brake pipe was connected. The assembly was mounted to the brake pipe storage bracket with the cap end of the tube protruding downwards to just above rail level. At the rear of the siding approximately three vehicle lengths from the buffer stops, a steel plate was mounted onto the sleepers, protruding upwards to just above rail level in line with the plastic pipe position on the last wagon. Using radios to communicate between test engineers at the rear of the siding and the locomotive driver, the train was propelled in the normal manner at slow speed into the siding. As the train approached the end of the siding, the plastic pipe contacted the metal plate, and as intended it fractured causing the brake pipe to vent and the train brakes to be applied from the rear. At this point the brake pip pressure would start to fall on the gauge in the locomotive cab; the driver then applied the train brake and the whole train came to a stand without any coupling shocks at all. The test was repeated a number of times, each time giving the same desired results, although we did use up the stock of plastic pipes we had prepared.

Fig.230 - 56039+56048 propelling loaded 102 tonne hoppers at Theale, March 1986

The tests highlighted the significant benefits of the AAR couplers, especially when propelling at Theale; and although the brake application device was proven to work effectively, the practicalities of using such a device operationally meant this was not implemented. The couplings on the PHA 102 tonne aluminium hopper wagons were subsequently modified with the installation of the AAR type couplers.

During February and March 1987 the testing section Freight Group undertook a series of impact tests on 2-axle open type wagons at the Derby station end of the RTC site, on the Way & Works sidings; the instrumented wagons were propelled at varying speeds into static load wagons to investigate axle-bearing damage and assessment of modified bearing arrangements. Acceleration and suspension displacement measurements were undertaken using various transducers fitted to the test wagons. These were coupled via an umbilical loom of instrument cables to the signal conditioning and recording equipment installed in Test Car 1 (TC1) which was parked on the adjacent line. The Class 08 diesel shunting locomotive that was used for this series of tests was number 08021. The locomotive was stabled at the RTC at the time following withdrawal from service, and was en-route from its former home depot at Leicester to Tyseley. I believe the locomotive still resides at Tyseley where it is preserved and currently painted in an early British Rail black livery, carrying its original number 13029 from when it was built in 1953.

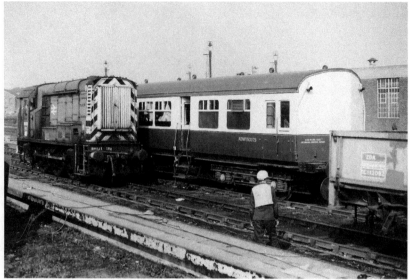

Fig.231 - Impact testing on the Way & Works sidings using 08021, February 1987

The test car was static during these tests, and was an example of such an occasion when the original central door extending steps of the auto-trailer came in very useful, because we had to climb in and out of the test car frequently during each day of testing. The weather was extremely cold, we even had a covering of snow for a few days, therefore were thankful for the hospitality of the Test Car, with its storage heaters and under seat electric heaters. The kitchen area in TC1 was also well appointed with a Baby-Belling cooker (no microwaves in those days) and the kettle was always on.

Four wagons were used during the tests to provide different weights for the stationary impact vehicle. Each wagon had its handbrake applied fully and wooden wheel chocks fitted under all wheels.

- DC112152, 2-axle wagon, in the tare condition, 17 tonnes,

- DC112055, 2-axle wagon, partly loaded with steel weights at 25 tonnes,

- DC112068, 2-axle wagon, loaded with steel to 50 tonne GLW,

- YNV Borail Wagon RDB948407, with plate frame bogies, loaded with rail stacks, 100 tonnes.

The test wagon was an ex OCA 2-axle wagon re-coded as a ZDA 'BASS' number DC112082; this vehicle was specially fitted with half-bore bearing to axlebox adaptors at both ends of one side of the wagon (corners 1 & 2) and full-bore bearing to axlebox adaptors at both ends of the other side (corners 2 & 4). For each set of tests the whole wagon was fitted with a different design of bearing such that a comparison of the effects of the shunting impacts on each bearing type installed in half and full-bore axlebox adaptors could be evaluated. The test instrumentation for these tests was more comprehensive that normally installed for wagon acceptance testing; this was in order to provide detailed information about the suspension movements and accelerations at the axleboxes to support the bearing investigations. At each corner of the wagon the longitudinal and lateral accelerations and displacements of the axlebox to bearing adaptors were recorded. Vehicle underframe longitudinal accelerations and buffer compression displacement measurements were also recorded during each test.

Fig.232 - Test Car 1 instrumentation area, OBA wagon impact testing, March 1987

Following completion of loading of the stationary impact wagons, the preparations progressed in the test car, including installation of the reel-to-reel magnetic tape recorders and UV paper trace recorders. The measurements of suspension clearances and instrumentation installation on the BASS test wagon then followed, and after completion of the signal conditioning set-up and calibration checks, the vehicles were moved to the Way & Works sidings. Test Car 1 was positioned on the No. 1 Way & Works siding line approximately 75 feet to the north of the impact test location.

Fig.233 - Test Car 1 Way & Works sidings during a very cold March 1987

The test wagon and Class 08 shunt locomotive were positioned on the No.2
Way & Works siding line to the north of the stationary impact ZDA and YNV
wagons. In order to accurately measure the speed of the test wagon, two wheel
flange operated micro-switches that were activated when the leading wheelset of
the test wagon passed over them, were installed on the inside edge of one of the
rails on No.2 Way & Works line, at a set distance a few metres apart. By
measuring the time between the activation of the two micro-switches, the exact
speed at which the test wagon was moving just before the impact could be
calculated. On the 25th February we connected the umbilical instrument cable
loom between the test wagon and TC1, and then carried out trial shunting runs to
re-check the instrumentation set-up, and fine tune the operations and
communications process to be used during the tests. The test process started with
the Class 08 drawing the test wagon towards the north end of the Way & Works
line, making sure not to go too far to damage the umbilical instrument cable loom.
The RTC yard shunting staff then uncoupled the wagon from the locomotive, and
when the tape data recorders were running, the Officer in Charge (OIC) gave
instruction via radio to the driver to start the test. The test wagon was propelled
up to the target test speed of 8 mile/h, and when the front on the locomotive
reached a point in line with the rear of TC1, the driver applied the locomotive
brake and stopped, allowing the test wagon to roll forward on its own to impact
with the stationary wagon. The shunting staff were on hand to apply the test
wagon handbrake following the impact to prevent the wagon from rolling back
towards the locomotive. Once confirmation had been received from the OIC that

all was in order with the test data collection, the locomotive was re-coupled to the test wagon to draw back to the north end of the line to start the next test.

Test ID	Date	Impact Vehicle	Test Vehicle / Condition
1	26th February	DC112152, 17 tonnes	DC112082 FAG Bearings (new design)
2		DC112055, 25 tonnes,	
3		DC112068, 50 tonnes	
4	2nd March	DC112152, 17 tonnes	DC112082 FAG Bearings (standard design)
5		DC112055, 25 tonnes,	
6		DC112068, 50 tonnes	
7		RDB948407, 100 tonnes	
8	4th March	DC112152, 17 tonnes	DC112082 SKF bearings
9		DC112055, 25 tonnes,	
10		DC112068, 50 tonnes	
11		RDB948407, 100 tonnes	
12	9th March	DC112152, 17 tonnes	DC112082 Timken bearings
13		DC112055, 25 tonnes,	
14		DC112068, 50 tonnes	
15		RDB948407, 100 tonnes	
16	12th March	DC112152, 17 tonnes	DC112082 FAG Bearings (standard design with reduced suspension clearances)
17		DC112055, 25 tonnes,	
18		DC112068, 50 tonnes	
19		RDB948407, 100 tonnes	
20		DC112152, 17 tonnes	

Table 22 - Bearing investigation impact test schedule

A total of twenty tests covering the four different bearing types were carried out between the 26th February and the 12th March 1987, with the BASS wagon returning to the EDU workshop between each test day to change the bearings.

The Class 08 locomotive driver was very consistent in propelling the wagon at the target speed, with the variance between tests of only about 0.5 mile/h; this made the analysis of the results much easier to compare between the different bearing types and impact wagon load configurations. In order to capture the detail of the acceleration and displacement data at the point of impact the chart paper trace recorder was run at a high speed, normally 250 mm of paper per second. The timing of the chart recorder operation was therefore crucial so as not to waste large quantities of the UV paper. During the tests I captured the point of impact with photographs and also on video. After the tests we were then able to play back the video at slow speed to scrutinise how the axleboxes moved at the point of impact in conjunction with the time history trace results of the acceleration and displacement transducer outputs.

Fig.234 - Impact testing bearings investigations - point of impact, March 1987

The results from the tests were collated into a formal report that was used to support the post-test inspection of each bearing and adaptor by the DM&EE metallurgy department, to investigate whether there were differences in any damage caused to the bearings during the tests between the bearing types or between the half and full bore adaptors. During the scrutiny of the video footage we noticed a phenomenon that although not related to the test remit, was nevertheless interesting; for a few seconds after the point of impact the wheelsets of the test wagon continued to rotate due to the inertia present, even after the test wagon had rebounded away from the impact wagon. This was more noticeable in the tests when the 100 tonne impact wagon was used, probably as a result of the slight different in buffer heights between the wagons.

The test section took advantage of the cold weather spell during early February 1988 to undertake some windscreen demister effectiveness tests on Class 58 and Class 87 locomotives.

Fig.235 - 87019 at Euston - windscreen demister test, February 1988.

These tests involved the windscreen glass supplier preparing two windscreens for both locomotive types by installing thermocouples (temperature measurement transducers) into a specially made version of the windscreen glass. The day before each test, the respective depot maintenance staff installed the special windscreens into the second man's side of one cab of the locomotives selected for test, numbers 87019 and 58023. We then connected the windscreen thermocouples, also measuring the cab and outside ambient temperatures, to a data logger unit installed in the cab of the locomotive. Each test run commenced in the very early hours of the morning with the locomotives operating on normal service runs whilst we rode in the cab to operate the data loggers, taking notes of the operations and ensuring the measurements were all in order. The test for the Class 87 was undertaken on the 12th February, starting at Longsight depot firstly working into Manchester Piccadilly with a train of empty coaching stock, then hauling a service train to Euston and back.

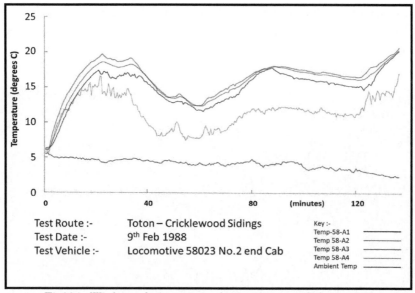

Test Route :-   Toton – Cricklewood Sidings

Test Date :-   9th Feb 1988

Test Vehicle :-   Locomotive 58023 No.2 end Cab

Key :-
Temp-58-A1
Temp 58-A2
Temp 58-A3
Temp 58-A4
Ambient Temp

Fig.236 - Windscreen demister test time history trace output, February 1988

Upon completion of the test run, the data logging equipment was removed and returned to Derby for analysis. Although the dedicated data analysis laboratory normally dealt with test output where data quantities were large and the analysis more complex, for tests such as the windscreen demister performance, it was normal for the test engineers to undertake the analysis themselves. In this case I undertook all the analysis and prepared the test report. The Class 58 locomotive test was conducted on the 9th February, operating from Toton Yard along the Midland Main Line to Cricklewood Recess Sidings via Manton and Corby, and then returning to Toton yard later the same morning.

During May 1988, Peter Metcalf and I set up a temporary instrumentation base in the back of the Austin Montego estate car to undertake hydraulic system pressure measurement tests at Carlisle Upperby Yard on a Cowans Sheldon Twin-Jib Crane / Tracklayer vehicle. The test instrumentation for this particular test consisted of a number of high range pressure transducers fitted into suitable test points in the hydraulic and air systems on both crane units of the tracklayer vehicle, a rack of purpose built amplifier / signal conditioning units, a UV paper trace recorder, and a TEAC 8-channel cassette data recorder. The equipment was powered by a small portable diesel generator that we took along with us in the car; this was placed outside away from the test area so as not to cause too much noise disturbance or interference to the test equipment. The tests consisted of operating both the tracklayer cranes simultaneously in varying movements, with varying loads attached to compare the operation of each crane unit. It was particularly important with this type of machine that the crane units at each end of the tracklayer operate in unison during all movements and under all load conditions.

Fig.237 - Twin Jib tracklayer at Carlisle Upperby, May 1988.

After completion of the tests and after return to Derby, the data was analysed to assess any differences between the two crane system operations and the findings passed to the crane owners to consider whether any repairs or modifications were required in order to maintain safe operation.

The Class 158 Diesel Multiple Units were particularly vulnerable during the leaf fall season to the build-up of wheel tread contamination, which caused poor track circuit actuation and a reduction in wheel and rail adhesion levels. Although the Class 158 units were fitted with track circuit assisting (TCA) devices the problems with track circuit actuation and low adhesion during braking persisted. A design of vehicle mounted mechanical wheel tread cleaning device was therefore proposed in the form of an Auxiliary Tread Brake (ATB) for fitment to a Class 158 unit for trial purposes. The British Rail Research department assessed the performance of the adhesion level improvements during a series of tests; however the suitability of the design of the installation in terms of integrity, performance and the effects on normal operation of the Class 158 DMU was the responsibility of the DM&EE department. I was tasked in March 1992 with leading the DM&EE work on the trial with a series of tests to confirm the integrity of the ATB bogie mounting bracket installation. The ATB brake block forces and the effects on the operational characteristics and passenger environment were also assessed when the ATB was operating. Class 158 unit number 158775 which had been operating in-service for a few years, was provided for the trials; the unit was delivered to the RTC Derby where the ATB modification work was carried out.

Fig.238 - Class 158 DMU 158775 fitted with ATB, March 1992

The instrumentation to record data from the transducers was installed within the saloon area of the one of the Class 158 vehicles prior to on-track trials along with a small diesel generator which was fitted in a specially built frame mounted under the inner headstock of one of the vehicles to provide power to the instrumentation systems, as well as a kettle.

Fig.239 - Class 158 Auxiliary Tread Brake Unit, March 1992

The assessment of the ATB brake block force was necessary to ensure that the brake force applied to the wheel tread was maintained during auxiliary tread brake operation. To achieve this, I measured the longitudinal loads transmitted through the actuator bracket with purpose designed, built and calibrated load cells. In order to confirm integrity of the brake actuator bracket and verify the design load case calculations it was proposed to measure the strains in the actuator bracket clamp bolts, the acceleration levels and the loads experienced by the bracket during dynamic operation of the auxiliary tread brake.

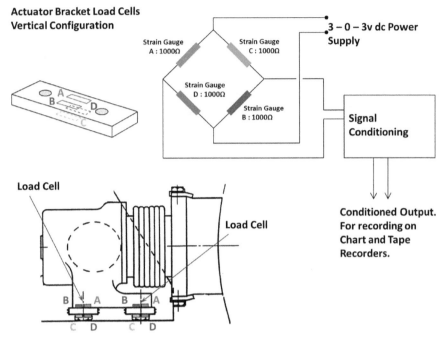

Fig.240 - Example actuator bracket load cell arrangement

The load cells on the actuator bracket were configured with 1000 Ω strain gauges connected in a Wheatstone full-bridge arrangement in order to obtain the optimum transducer output level. The load cell was calibrated during static tests during which known vertical, lateral an longitudinal forces were applied to the actuator bracket and the ATB brake block whilst recording the load cell output.

The dynamic test runs were undertaken in March 1992 at the Old Dalby test facility and also whilst running over a typical route on which the Class 158 units were operating in service, between Derby, Liverpool and Norwich. In order to demonstrate that the normal operation of the Class 158 units was not compromised, measurement of the riding properties was undertaken using body acceleration and suspension displacement transducers. The data was then analysed to show any differences in the dynamic performance during ATB operation. Comparison assessment was carried out using time history, peak acceleration and frequency content analysis. The analysis concluded that the operation of the auxiliary tread brake did not detrimentally affect the ride accelerations, suspension operation, or the passenger comfort levels. To assess the effect of the auxiliary tread brake on the overall brake stopping distance performance on the Class 158 unit, tests were conducted with and without the auxiliary tread brake operative. The brake stopping distance tests were carried out at the Old Dalby test facility, whilst this line was eminently suitable for carrying out dynamic tests off the main railway network, it does not have any level track sections. It was therefore necessary to carry out brake stopping distance tests on both rising and falling gradients and analyse the results to confirm the level track brake stopping distance performance. The brake performance on level track was then plotted on a graph to confirm the effects of the auxiliary tread brake on normal service train braking performance of the Class 158 unit. It can be seen that as expected, the overall stopping distance was shorter when the auxiliary tread brake was operative. At the maximum speed tested the average deceleration rate of a full service brake application was increased by approximately 0.25%. When considering the improvements in adhesion levels that the auxiliary tread brake was expected to give, the small increase in brake performance was likely to be discernible by drivers.

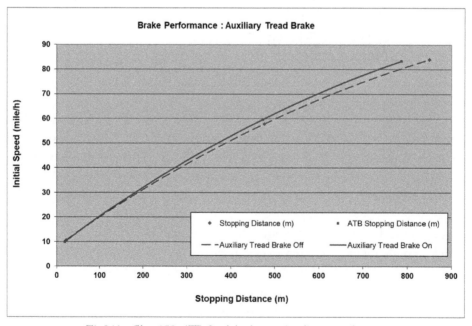

Fig.241 - Class 158 ATB fitted, brake stopping distance performance

The interior noise levels were measured using a proprietary sound pressure level (SPL) meter in the Class 158 passenger saloon area, above the power bogie to which the auxiliary tread brake was fitted. The SPL meter was set to record 'A' weighted average sound pressure levels; this being the most suitable weighting for lower level sound signal recording. Readings were taken when the unit was running over sections of both welded and jointed rail types, during braking, with and without the auxiliary tread brake operative. The noise levels recorded showed a marginal increase in sound pressure level with the auxiliary tread brake in operation. This was expected as the disc braking system fitted on the Class 158 was inherently quieter in operation than a block braked system. Wheel tread temperatures were also measured and assessed during the trials to ensure that the operation of the auxiliary tread brake did not promote wheel tread damage, and to ensure that excessive heat was not being generated by the operation of the auxiliary tread brake. The temperatures were measured using a rubbing thermocouple positioned approximately 300mm from the actuator block on the centre of the wheel tread on one wheel. It was noted that there was some interference on the wheel tread temperature measurement that was typical of the plunger type transducer used. Subsequently the wheel tread temperature data was filtered at a low frequency using a 2 Hz low-pass filter to remove the interference. In addition to the main line running, spot checks of the wheel tread temperatures and also the auxiliary tread brake block temperatures was taken during the braking tests at the Old Dalby facility. Although the block temperatures recorded were relatively high, these levels were not considered to be uncharacteristic for a cast iron brake block fitted vehicle. An example of the time based cycle counting analysis of the wheel tread temperature records is shown in Fig.242. The level resolution for the analysis was selected at 2 degrees and plotted on a linear distribution of time that the temperature was recorded within each level. The maximum wheel tread temperature being 199 degrees C and the mean temperature for the whole test run was 80.1 degrees C.

A final test was conducted with the aim of simulating a failure scenario, whereby the ATB could not be released and remained in contact with the wheel. The test was carried out at the Old Dalby line, with the auxiliary tread brake fully applied whilst driving the Class 158 unit for approximately five miles at speeds up to 40 mile/h. The temperature levels during the failure mode tests did not exceed those seen during normal operation tests on the main line.

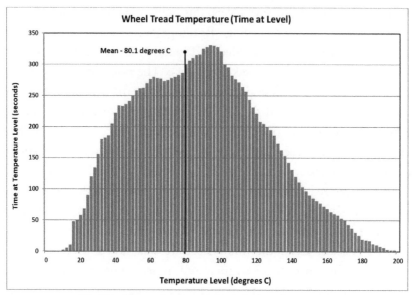

Fig.242 - Time at Level analysis of wheel tread temperature

Overall the results of the tests auxiliary tread brake device on the Class 158 unit demonstrated that the system did not have any detrimental effect on the normal operation of the units, and combined with the improved wheel tread condition prevalent as a result of the installation of the auxiliary tread brake progress with a production fit of such a system to Class 158 units was authorised.

SPECIAL TESTS AND EVENTS

MEGATRAIN

During 1991 whilst working for the DM&EE Testing Section at Derby, I was lucky enough to be part of the test team that carried out various tests during the Foster Yeoman 12000 tonne 'Megatrain' trial. On 24th May the team of four engineers including myself headed to Merehead and prepared three Class 59 locomotives and two wagons with test instrumentation and data recording equipment in readiness for the test on the Sunday 26th May 1991 which was formed of 115 wagons with additional locomotives in the centre and at the rear of the formation. The test instrumentation included fitting of brake pipe pressure and brake cylinder measurement transducers to each locomotive, and three screw-link couplings that had been fitted with strain gauges and calibrated at the RTC in Derby prior to the tests, in order to record the drawgear loads between the leading and centre locomotives and the adjacent wagons. Radio telemetry units were installed in all three locomotives to enable synchronisation between the data recording equipment in each locomotive.

Fig.243 - 59005 at Merehead during preparations for the 12000t Megatrain trial

The test equipment, including the signal conditioning amplifiers, paper chart recorders and VHS data recorders that were secured in the locomotive cabs were powered by using inverter units installed in the engine compartments of 59005 and 59001. The inverters were required to convert the 74 volts DC main battery supply to the 240 volts AC power required for the instrumentation.

THEALE (TOPS Destination 74706)				CRAWLEY (TOPS Destination 88002)		
1	OK19317	JHA		41	OK3268	JYA
2	OK19390	JHA		42	OK3290	JYA
3	OK19391	JHA		43	PR26481	JUA
4	OK19363	JHA		44	PR26465	JUA
5	OK19358	JHA		45	PR26471	JUA
6	OK19324	JHA		46	OK3275	JYA
7	OK19350	JHA		47	OK3292	JYA
8	OK19380	JHA		48	PR26541	JUA
9	OK19347	JHA		49	PR26554	JTA
10	OK19312	JHA		50	OK3270	JYA
11	OK19307	JHA		51	PR26473	JUA
12	OK19343	JHA		52	PR26501	JUA
13	OK19383	JHA		53	PR26461	JUA
14	OK19310	JHA		54	OK3299	JYA
15	OK19303	JHA		55	PR26462	JUA
16	OK19345	JHA		56	PR26536	JUA
17	OK19332	JHA		57	PR26513	JUA
18	OK19398	JHA		58	OK3273	JYA
19	OK19320	JHA				
20	OK19369	JHA				
21	OK19364	JHA				
22	OK19377	JHA				
23	OK19365	JHA				
24	OK19340	JHA				
25	OK19325	JHA				
26	OK19394	JHA				
27	OK19328	JHA				
28	OK19393	JHA				
29	OK19309	JHA				
30	OK19318	JHA				
31	OK19386	JHA				
32	OK19382	JHA				
33	OK19327	JHA				
34	OK19395	JHA				
35	OK19341	JHA				
36	OK19397	JHA				
37	OK19331	JHA				
38	OK19368	JHA				
39	OK19346	JHA				
40	OK19308	JHA				

Table 23 - Megatrain formation front portion, 58 wagons headed by 59005

The test train comprised 115 loaded 102 tonne GLW aggregate wagons for five different destinations that were combined together for the trial run between Witham and Theale. The majority of the formation had the centre knuckle couplers within the wagon rakes of each section, and screw-link couplings at the ends between each section and the locomotives.

ACTON (TOPS Destination 73238)		
59	OK19313	JHA
60	OK19354	JHA
61	OK19396	JHA
62	OK19302	JHA
63	OK19301	JHA
64	OK19384	JHA
65	OK19375	JHA
66	OK19351	JHA
67	OK19336	JHA
68	OK19370	JHA
69	OK19374	JHA
70	OK19373	JHA
71	OK19319	JHA
72	OK19314	JHA
73	OK19315	JHA
74	OK19316	JHA
75	OK19335	JHA
76	OK19344	JHA
77	OK19359	JHA
78	OK19305	JHA

BOTLEY (TOPS Destination 86203)		
97	OK19304	JHA
98	OK19321	JHA
99	OK19362	JHA
100	OK19349	JHA
101	OK19330	JHA
102	OK19371	JHA
103	OK19357	JHA
104	OK19352	JHA
105	OK19355	JHA
106	OK19334	JHA
107	OK19326	JHA
108	OK19360	JHA
109	OK19381	JHA
110	OK19372	JHA
111	OK19388	JHA
112	OK19392	JHA
113	OK19329	JHA
114	OK19323	JHA
115	OK19311 #	JHA

WOKING/SALISBURY (TOPS destination 86036/86121)		
79	PR26548	JTA
80	OK3277	JYA
81	OK3274	JYA
82	PR26503	JUA
83	OK3321	JYA
84	OK3287	JYA
85	PR26476	JUA
86	OK3320	JYA
87	OK3283	JYA
88	PR26550	JTA
89	OK3327	JYA
90	OK3300	JYA
91	OK3302	JYA
92	PR26515	JUA
93	PR26522	JUA
94	PR26478	JUA
95	PR26508	JUA
96	PR26563	JTA

Note # - Wagon OK19311 was replaced with an O&K Box Wagon at Merehead before departure, no record kept of the replacement wagon number.

Table 24 - Megatrain rear portion, 57 wagons with 59001 between the front and rear portions

On the day of the test the DM&EE team arrived in the early evening to make final preparations to the test instrumentation. We all then took advantage of a break for a hot meal whilst the final marshalling of the trains was carried out to get all the wagons and locomotives in the correct positions. The team returned mid-evening to connect the transducer cables between the leading wagon OK19317 and the rear of 59005, and also between the 59th wagon, OK19313 and the rear of 59001. The normal route for the departure train movements at Merehead was by propelling the train from the sidings to Merehead Quarry Loop West Junction, then heading on their way past Merehead Quarry Loop East Junction and along the branch line to Witham East Somerset Junction.

Fig.244 - The author (right) preparing instrumentation on 59001 [Hugh Searle]

The normal arrival of trains came in direct from Merehead Quarry Loop East Junction down the gradient into the sidings. Due to operational constraints of the train length the route to get the trial train out of the Merehead Quarry sidings needed to be via the normal arrival line direct to Merehead Quarry Loop East Junction, however the gradient on the branch meant introducing high risk to the movement therefore the train was formed in Merehead Quarry sidings as two separate trains, each approaching 6000 tonnes. The first portion was headed by 59005 with assistance provided by 59003 + 59002 at the rear, banking the train out of the quarry to Witham East Somerset Junction. The banking locomotives 59003 + 59002 then returned back along the branch to the quarry sidings to assist in banking the second portion. It took two attempts to move the second portion of the train headed by 59001 out of the sidings. With 59003 + 59002 powering hard to get the train up the gradient, as the rear of the train traversed a set of points at the Torr works end of the quarry sidings, the front buffer of 59003 overrode the rear wagon buffer and became locked, damaging and derailing the rear bogie of the rear wagon OK19311 and removing one of 59003s buffers completely. The wagon was subsequently re-railed and replaced with an O&K Box wagon, and thanks to the sterling efforts of the Foster Yeoman locomotive maintenance crew the buffer on 59003 was replaced; the incident delaying the proceedings by about four hours. 59003 + 59002 eventually being re-coupled to the rear of the second portion of the train and the ensemble successfully headed off to Witham East Somerset Junction. It took about 30 minutes to traverse the four mile branch line and at about 4am the two trains comprising of the 115 wagons of JHA, JYA, JUA and JTA types totalling 11982 tonnes were coupled together to make one 5415 ft long train with 59005 at the head, 59001 in the centre and 59003 at the rear. We coupled the instrument cables between the leading end of 59001 and the 58th wagon OK3273, and with test engineers located in the three locomotive cabs, recorded the brake pipe and brake cylinder pressures whilst a statutory brake test was conducted by the traincrew. I was the test engineer in charge of the test equipment in 59001.

Fig.245 - 59001 in the centre of the 12000t Megatrain [Hugh Searle]

The whole formation then operated as head code 7Z09 from Witham East Somerset Junction and was powered up to about 30 mile/h by 59005 (with a little assistance in power from 59001 in the centre of the train). After about two miles an initial position brake application was made that bought the train speed down to around 20 mile/h and as the train brake were released high coupling tensile loads were recorded above the 560 kN normal working load limits for screw couplings.

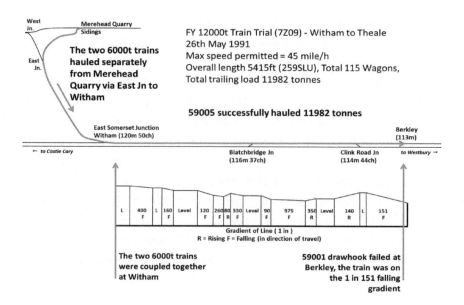

Fig.246 - 12000t Megatrain trial route and gradient profile representation

Power was re-applied on both 59005 and 59001 and the train speed increased steadily to 35 mile/h for about three miles before power was shut off on both locomotives and the train coasted past Clink Road Junction. The brake was then applied again to initial position from 59005 bringing the speed down to about 20 mile/h near 113 milepost at Berkley; as the brakes were in the process of being released, a longitudinal shock occurred through the centre portion of the train causing coupling loads to increase further above the 560kN limit. This resulted in the drawhook at the leading end of 59001, which was coupled the 58th wagon (OK3273), fracturing allowing the train to part. The air brake pipes parted causing an emergency brake application and the two halves of the train came to rest about 100 metres apart. The test was subsequently terminated at Berkley due to the time taken to effect repairs, and the two parts of the train made their way separately, initially to Westbury where the test team removed the test equipment before returning to Derby along with the fractured drawhook, which was presented to the DM&EE metallurgy department for analysis.

Fig.247 - 59001 following drawhook failure [Hugh Searle]

I also analysed the drawbar force data recorded from the leading and trailing end instrumented couplings on 59001; the peak tensile coupling load measured in the couplings at both ends of the locomotive was in the order of 840 kN at the point of the drawhook failure. The train speed at the point of the failure was 16 mile/h, the whole train was on the 1-in-151 falling gradient, the brakes at the front of the train were fully off and the brakes at the rear of the train were still releasing. The engineering standards in place at the time required locomotive drawhooks to be capable of withstanding a Proof Force of 700 kN (the maximum force expected in normal service) and an Ultimate Force of 1200 kN (the force the coupling should be able to bear without failing), therefore although the drawhook had been subjected to operational loads greater than those expected in normal service, quite clearly it should not have failed. The metallurgists conducting the analysis of the drawhook were able to assist with a possible contributory factor, when they found a small flaw in the casting in the cup of the hook.

Fig.248 - 59001 failed drawhook [Hugh Searle]

It was the opinion within the test team that this contributed to the failure of the drawhook; however analysis of the brake pressure data also provided details of the state of the brake application at the different points in the train at the time of the failure. It was also clear that the operation of such a long train over varying gradients contributed to the greater than expected normal service coupling loads. It might well have been a different story had the Class 59 locomotives been fitted a Tightlock knuckle type or Swing-head coupler such as those installed on some of the later Class 66 type locomotives.

IC225 DEMONSTRATION

Having been involved in the Mk IV coach project from the early stages of the SIG bogie development through to the final tests before introduction into service, I was pleased to receive an invite from the Managing Director of InterCity to attend a demonstration non-stop run of the InterCity 225 on the 26th September 1991 between Kings Cross and Edinburgh.

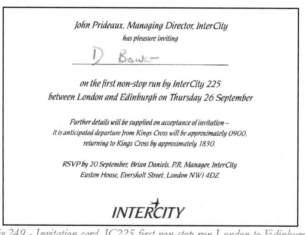

Fig.249 - Invitation card, IC225 first non-stop run London to Edinburgh

A special shortened Mk IV formation with Class 91 locomotive 91012 hauling five Mk IV coaches plus a Mk IV DVT was provided for the run. Five hundred specially invited guests from the railway industry and the press were on board for the 09:00 departure from Kings Cross running as 1T01, non-stop to Edinburgh Waverley in 3 hours and 29 minutes, averaging a speed of 112.8 mile/h. After arrival at Edinburgh, the guests, myself included, enjoyed a reception at the Balmoral Hotel on Princes Street just outside Waverley station, where the high speed run was declared a record shortest time for a train to cover the East Coast Main Line route from Kings Cross. The return trip home travelling by normal service train back to Derby was somewhat less exciting, and took over an hour longer than the high speed journey north from London.

Fig.250 - Invitation card, IC225 crosses Durham Viaduct by Terence Cuneo

PETERBOROUGH 150

The testing team were called upon in mid-1995 to assist with data recording to confirm running stability during some high speed runs on the East Coast Main Line that were to coincide with the Peterborough 150 celebrations on the 2nd June. Preparations for the tests were carried out at Bounds Green depot where the train was formed of five Mk IV coaches, a DVT and a Class 91 locomotive. The vehicles in the formation being 91031 + 12204 + 11273 + 10323 + 11274 + 11275 + 82231, with the Class 91 at the north end. One of the challenges we faced with this particular set-up was that the visibility of any test instrumentation cables or equipment had to be kept to an absolute minimum because there were many visitors invited to board the train on the 2nd June special high speed run. A tachometer was fitted to one axle end of the TOE vehicle 12204 where the instrumentation base was located. Signal conditioning amplifiers were connected to the bogie frame mounted lateral accelerometers located on the Class 91, the TOE 12204 and the DVT 82231. The associated instrumentation cables were routed on the outside of the train along the underframes of each vehicle to connect the transducers, and also to connect the intercom system and digital speed displays fitted in both driving cabs to the instrument area of the TOE.

As part of the event the train was to be filmed from the air, therefore in order to make it easy for the film crew in the helicopter to distinguish the high speed test train from other trains, the buffer heads of the Class 91 and the DVT were painted with white crosses.

Fig.251 - The author during testing on the Peterborough 150 trial run, June 1995

Preparations and formation of the train were completed by the 30th May at Bounds Green. The train was then moved on a transit and instrument proving checks run on the 31st May from Bounds Green to Heaton depot near Newcastle. During this run we initially set up and calibrated the tachometer output, speed and distance measurement system and the speed displays in the cabs, following which a number of braking tests were conducted at speeds up to 125 mile/h to confirm the shorter formation of the train did not affect the brake stopping distance performance. A rehearsal for the high speed run was carried out between Newcastle and Bounds Green depot on the 1st June, operating at speeds up to 153 mile/h between Darlington and York. The train was then returned to Heaton depot where it was cleaned in preparation for the next day's event as part of the Peterborough 150 celebrations.

The high speed run with specially invited guests and members of the press on board ran as planned on 2nd June operating from Newcastle to Peterborough including running for more than 14 miles at speeds above 150 mile/h and attained a maximum of speed of 154 mile/h between Darlington and York. A speed of 154 mile/h was attained a second time for a short distance between Grantham and Peterborough. The train then made a short stop at Peterborough before continuing at normal line speeds into London Kings Cross. No visitors were allowed in the TOE vehicle during the high speed run which meant we could continue undisturbed monitoring the instrumentation and bogie stability during the run. After arrival at Kings Cross the train was moved back to Bounds Green depot where we removed the test equipment and instrument cables.

Fig.252 - The author with Mick Wright at Peterborough, June 1995

BIBLIOGRAPHY

- Amberley Books, John Dedman, British Rail Wagons 1980 -2015

- British Rail DM&EE (Dec 1982), "Proposed Criteria and Procedures to be used for the Ride Acceptance Testing of Freight Vehicles"

- British Rail DM&EE (Sept 1989), "Testing Section Bulletin No.88 - MkV Amplifier Manual"

- D.M.Bower (2006) IMechE Treatise Paper, "Prototype Auxiliary Tread Brake for Class 158 Diesel Multiple Units"

- G.H.Ryder (1983). "Strength of Materials, Third Edition"

- Guild Publishing, British Locomotives of the 20th Century, O.S.Nock, 1985

- Ian Allan ABC-British Rail Motive Power Combined Volume 1989

- Ian Allan, Modern Railways Special, Derby Railway Technical Centre, 1989

- Jane's Railway Special, Geoffrey Freeman Allen, The Yeoman 59s

- NPT Publishing, On-Track Plant 2003

- Oxford Publishing Co, Rail Atlas of Britain (Issue 3) 1980

ABBREVIATIONS

AAR	Association of American Railroads
ARC	Amey Roadstone Corporation
ATB	Auxiliary Tread Brake
BRB	British Railway Board
BREL	British Rail Engineering Limited
BRR	British Rail Research Division
BSK	Brake Standard Corridor (Passenger Coach Type)
CM&EE	Chief Mechanical & Electrical Engineers (Regional)
DM&EE	Director of Mechanical & Electrical Engineering (British Rail)
DMU	Diesel Multiple Unit
D of CE	Director of Civil Engineering
EDU	Engineering Development Unit (RTC DM&EE Workshop)
EMC	Electro-magnetic Compatibility
EMU	Electric Multiple Unit
ETH	Electric Train Heating
FK	First Corridor (Passenger Coach Type)
GLW	Gross Laden Weight
HOBC	High Output Ballast Cleaner
HP	Hewlett Packard
HSFV1	High Speed Freight Vehicle 1 (Suspension development wagon)
HST	High Speed Train (Inter-City 125)
HSTRC	High Speed Track Recording Coach
ISO	International Organization for Standardisation
LWRT	Long Welded Rail Train
MENTOR	Mobile Electrical Network Testing, Observation and Recording
MPV	Multi-Purpose Vehicle
NER	North Eastern Railway
NMT	Network Rail New Measurement Train
OBA	On-Board Data Analysis System
OIC	Officer in Charge
PCZC	Peak Count Zero Crossing (Data Analysis Method)

PO	Pullman Open (Passenger Coach Type)
RCE	Regional Civil Engineers department
RFI	Radio Frequency Interference
RSSB	Rail Safety & Standards Board
RTC	Railway Technical Centre (Derby)
SR	Southern Region
SPL	Sound Pressure Level
SV	Service Vehicle (Passenger Coach Type)
TO	Tourist Open (Passenger Coach Type)
TOD	Tourist Open Disabled (Passenger Coach Type)
TOE	Tourist Open End (Passenger Coach Type)
TC1	Test Car 1 (ADW150375)
TC2	Test Car 2 (ADB975397)
TC6	Test Car 6 – (ADB975290)
TC10	Test Car 10 – (ADB975814)
TGS	Trailer Guard Standard(Passenger Coach Type)
TOPS	Total Operations Processing System
TS	Trailer Standard (Passenger Coach Type)
UIC	International Union of Railways
VAB	Vehicle Acceptance Body

APPENDICES

1. SUMMARY OF TESTING CAREER

This book has taken a look back at the wide range of traction and rolling stock testing experiences I was lucky enough to encounter during the 1980s and 1990s within the hugely varied and interesting field of rail vehicle testing. The roles that I undertook during this period were:-

Date	Organisation	Grade
1980 - 1984	British Railways Board (BRB), DM&EE	Trainee Technician Engineer
1984 - 1985	BRB, DM&EE Freight Locomotive Design Office	Technical Officer
1985 - 1988	BRB, DM&EE Vehicle Testing Section	Technical Officer
1988 - 1991	BRB, DM&EE Vehicle Testing Section	Senior Technical Officer
1991 - 1997	BRB, DM&EE Vehicle Testing Section	Senior Engineer
1997 - 2000	Serco Railtest, Testing Services	Project Leader

Table 25 - Career Summary 1980 to 2000

The tables in these appendices form a detailed account taken from my diaries logged between 1983 and 1999, providing some more history and information for those wishing to peruse the list of tests that I was on board whilst undertaking tests or as an Officer in Charge of the tests.

Where I did keep the records of the locomotive numbers and test coach used on each test, these have been logged for reference.

These appendix tables and the summary tables below provide insight into the massive variety of experience throughout the 1980s and 1990s that I was fortunate enough to gain within the field of railway vehicle testing. I was directly involved with over 200 rail vehicle testing projects, I authored 147 technical reports relating to railway vehicles testing projects undertaken, and issued 42 Certificates of Conformance for Acceptance Testing as a Railtrack Authorised signatory.

Note that following revisions to the rolling stock acceptance process that was implemented at privatisation, the responsibility for test certification moved across to the Vehicle Acceptance Body (VAB) organisations in 1999, to which formal approved reports were subsequently issued.

Summary by Role	Test Projects	Technical Test Reports	Report Approval	Railtrack Certificates
1983 - 1985	2	-	-	-
1985 - 1988	38	28	-	-
1988 - 1991	21	32	-	-
1991 - 1997	69	70	-	2
1997 - 2000	73	17	49	40
Totals	203	147	49	42

Table 26 - Summary of Experience by Role

Summary by Vehicle Type	Test Projects	Test Reports	Report Approval	Railtrack Certificates
Multiple Units	10	9	-	-
Locomotives	12	7	-	-
Coaching Stock	10	23	-	-
Freight Vehicles	114	75	42	26
On-Track Plant	40	16	7	16
Road Rail	4	1	-	-
Other	13	16	-	-
Totals	203	147	49	42

Table 27 - Summary of Experience by Vehicle Type

Year	Test Car 1 [TC1] ADW150375	Test Car 2 [TC2] ADB975397	Test Car 3 [MENTOR] ADB975091	Test Car 6 [TC6] ADB975290	Test Car 10 [TC10] ADB975814	Other Dynamic Tests not using DM&EE Test Cars
1983	-	-	-	16	-	3
1984	-	-	-	-	-	1
1985	11	9	-	-	-	22
1986	24	27	3	4	-	9
1987	19	9	-	-	-	24
1988	15	19	-	6	-	6
1989	21	5	-	-	18	21
1990	21	7	-	-	5	37
1991	31	3	1	-	-	17
1992	17	5	-	-	-	13
1993	21	12	-	-	-	23
1994	23	12	-	-	-	16

Year	Test Car 1 [TC1] ADW150375	Test Car 2 [TC2] ADB975397	Test Car 3 [MENTOR] ADB975091	Test Car 6 [TC6] ADB975290	Test Car 10 [TC10] ADB975814	Other Dynamic Tests not using DM&EE Test Cars
1995	8	7	-	-	-	29
1996	4	3	-	-	-	3
1997	9	2	-	5	-	8
1998	-	10	-	22	-	1
1999	-	10	-	16	-	4
Totals	224	140	4	69	23	237

Table 28 - Summary of Test Runs per annum by Test Car

The table above is extracted from my diary logs and shows the extent of the dynamic test runs per annum that I was on-board; totalling nearly 700 occasions between 1983 and 1999, a large proportion of which were undertaken on Test Car 1 and Test Car 2.

Fig.253 - Test Cars 1 & 2 at Cricklewood, November 1986

2. TESTING DIARY - 1983

Diary 1983	Test Description	Loco	Test Car
Feb 15	Class 58 Static torsional stiffness (ΔQ/Q) test	58001	
Feb 16	Class 58 Bogie rotational resistance test and parking brake tests at the EDU	58001	
Feb 18	Derby - Bedford Class 58 ride & stability assessment hauled by a Class 45 locomotive coupled to TC6 and two coaches	58001 & Class 45	[TC6]
Mar 22	Class 58 wheel load equalisation tests, RTC Weighbridge	58001	-
Aug 2	Class 58 Exhaust Noise Levels, Doncaster Works	58004	-
Aug 3	Class 58 Traction Tests up to 60 mile/h between Toton and Cricklewood, coupled with TC6 + 36 loaded HDA Wagons (total 1656 tonnes)	58001	[TC6]
Aug 4	Class 58 Traction Tests up to 60 mile/h between Toton and Cricklewood, coupled with TC6 + 42 loaded HDA Wagons (total 1926 tonnes)	58001	[TC6]
Aug 5	Class 58 Traction Tests up to 60 mile/h between Toton and Cricklewood, coupled with TC6 + 42 loaded HDA Wagons (total 1926 tonnes)	58001	[TC6]
Aug 8	Class 58 Traction Tests up to 80 mile/h between Derby and Cricklewood coupled with TC6 + 8 Coaches + Class 25 locomotive (total 370 tonnes)	58001 & Class 25	[TC6]
Aug 9	Class 58 Traction Tests up to 80 mile/h between Derby and Cricklewood coupled with TC6 + 8 Coaches + Class 25 locomotive (total 370 tonnes)	58001 & Class 25	[TC6]
Aug 11	Class 58 Traction Tests up to 80 mile/h between Derby and Cricklewood coupled with TC6 + 8 Coaches + Class 25 locomotive (total 370 tonnes)	58001 & Class 25	[TC6]
Aug 12	Class 58 Traction Tests up to 80 mile/h between Derby and Cricklewood coupled with TC6 + 8 Coaches + Class 25 locomotive (total 370 tonnes)	58001 & Class 25	[TC6]
	Class 58 curving tests within Derby Litchurch Lane Works	58001	[TC6]
Aug 24	Class 58 Ride, Stress tests between Derby and Cricklewood with coaching stock, tests up to 90 mile/h (58001 + TC6 + RDB975422 + RDB975428)	58001	[TC6]
Aug 25	Class 58 Ride, Stress tests between Derby and Cricklewood with coaching stock, tests up to 90 mile/h (58001 + TC6 + RDB975422 + RDB975428)	58001	[TC6]
Aug 26	Class 58 Ride, Stress tests between Toton and Cricklewood with 46 loaded HDA Wagons (total trailing load 2232 tonnes)	58001	[TC6]

Diary 1983	Test Description	Loco	Test Car
Aug 31	Class 58 curving tests Burton-on-Trent and Branston Junction triangle	58001	[TC6]
Sep 19	Class 58 Bogie Stress tests between Derby and Bedford coupled with TC6 + TC10 + RDB975136, tests up to 90 mile/h	58001	[TC6]
Sep20	Class 58 Bogie Stress tests between Derby and Cricklewood coupled with TC6 + TC10 + RDB975136, tests up to 90 mile/h	58001	[TC6]
Sep 21	Class 58 Bogie Stress tests between Toton and Cricklewood coupled with 40 loaded HDA wagons (total trailing load 2061 tonnes)	58001	[TC6]
Sep 26	Class 58 Light locomotive braking tests up to 90 mile/h between Crewe and Winsford	58001	-
Sep 27	Class 58 Light locomotive braking tests up to 90 mile/h between Crewe and Winsford	58001	-
Sep 28	Class 58 Static brake tests at Burton-on-Trent Wagon depot coupled with TC6 + 40 HAA wagons	58001	-

3. TESTING DIARY - 1984

Diary 1984	Test Description	Loco	Test Car
May 18	2 x Class 56 hauling a 4600 train of loaded PTA aggregate wagons for a performance testing.	2 x Class 56	-

4. TESTING DIARY - 1985

After completing my 5 year British Rail apprenticeship, I was lucky enough to apply for and secure a post as a Technical Officer on the DM&EE Testing Section, in the Freight & On-Track Plant team, based on the top floor of Derwent House at the Railway Technical Centre in Derby.

Test Logs 1985	Loco	Test Car
During 1985, my test diary records were less thorough than in later years, therefore specific dates for some tests are not available. Where records of test dates from 1985 are available these are noted in the lower part of this table.		
Speno URR 16P-46 Switch and Crossing Grinder static tests	-	-
Plasser GWM110 - DR79401 and Joint Straightener DR86100 Train- Acceptance- ride performance coupled with TC1 up to 45 mile/h on Midland Main line route.	-	[TC1]
51 tonne GLW Railease Scrap Steel Box Wagon, static tests, (ΔQ/Q)	-	-
80 tonne sodium tank by standard wagon for Octel, static tests, (ΔQ/Q) and bogie rotation	-	-
Esso LPG Bogie Tank Wagon TCA78036, static bogie rotation and (ΔQ/Q)tests - at the RTC Derby	-	-
Plasser TASC45 -DX68500 self-powered ride comfort and braking tests using portable data recording equipment	-	-
Marcon aggregate wagon MAR17713, static bogie rotational resistance measurements following derailments and excessive flange wear reports	-	-
Railease PFA Bogie Container Flat Wagon - RLS95477, Slip/brake testing, Crewe - Winsford	-	[TC2]
Permaquip Broken Rail Emergency Vehicle - static (ΔQ/Q) test	-	-
Long Welded Rail Train power wagon DR89001, Ride Test, Derby - Bedford – Derby	-	[TC1]
51 tonne GLW Railease box wagon, static tests including weighing and torsional stiffness (ΔQ/Q) tests	-	-
SUKO Oil Tank 65818 with UIC Tyre Profile - Ride Tests, Tare and Laden conditions.	-	[TC1]
Plasser 08-16 Universal Tamper/Liner DR73801, static torsional stiffness (ΔQ/Q) and bogie rotation tests and braking tests self-powered	-	-
PFA Bogie Container Flat Wagon(PFA) GMC92542 fitted with a modified brake system including tandem brake cylinders, slip/brake tests up to 75 mile/h between Crewe - Winsford	-	[TC2]
URR48/4 Speno Rail Grinding train DR79211-79214, self-powered ride and braking performance tests on the Midland Main Line route	-	-

Test Logs 1985	Loco	Test Car
80 tonne container flat wagon Avon Container Flat (fitted with VNH-1 Bogies) AVON92563 - static brake tests including parking brake, Slip/brake performance tests up to 75 mile/h - Crewe - Winsford, Tare and Laden conditions	-	[TC2]
46 tonne POA Aggregate / Scrap Metal Wagon TRL5142, this was the first of a series of wagons built on old TTA tank wagon underframes, the tests involved static torsional stiffness ($\Delta Q/Q$) and ride performance tests on the Midland Mail line route to Bedford	-	[TC1]
GMC refuse PFA Bogie Container Flat Wagon GMC92528 acceptance, braking performance, slip/brake tests up to 75 mile/h between Crewe and Winsford	-	[TC2]
Structure Gauging Train - measurement car ride evaluation using portable test equipment	-	-

Diary 1985	Test Description	Loco	Test Car
Apr 15	Procor Prologie Wagon PR95300, Ride Test, Tare - Derby - Syston Junction - Derby	Class 47	[TC1]
Apr 25	Procor Prologie Wagon PR95300, Ride Test, Laden - Derby - Melton Mowbray - Derby	Class 47	[TC1]
June 16	Procor Prologie Wagon PR95300 Tare, bogie rotational resistance retest with yaw dampers fitted	-	-
June 17	Procor Prologie Wagon PR95300, Ride Test, Tare, retest with yaw dampers fitted - Derby - Kettering - Derby	Class 47	[TC1]
June 20	Procor Prologie Wagon PR95300, Ride Test, Laden, retest with yaw dampers fitted - Derby - Kettering - Derby	Class 47	[TC1]
June 23	Slip/brake Tests between York - Thirsk on ECML - Mk III Coach (W44029) with Lucas Girling axle mounted disc brakes fitted with Ferodo brake pads, testing to 95 mile/h maximum	Class 47	[TC2]
June 28	Class 31/4 rough ride investigations, Doncaster - Peterborough - Lincoln - Doncaster (standard suspension)	31467	-
June 30	Slip/brake Tests between York - Thirsk on ECML - Mk III Coach (W42257) with modified BT10 bogies and Knorr Bremse axle mounted disc brakes fitted with Becorit brake pads, testing to 95mile/h maximum	47221	[TC2]
July 2	Foster Yeoman 2550 tonne stone train - Merehead - Westbury - Theale, coupling load tests	56001 + 56050	-
July 3	Foster Yeoman 2550 tonne stone train coupling load tests at Theale stone terminal	56039	-
July 4	Foster Yeoman 2550 tonne stone train coupling load tests Theale - Westbury	56043	-

Diary 1985	Test Description	Loco	Test Car
July 15	Class 31/4 rough ride investigations, York - Scarborough - York (standard suspension)	31419	-
July 16	Class 31/4 rough ride investigations, Doncaster - Peterborough - Lincoln - Doncaster (modified suspension)	31467	-
July 22	HST Braking tests on Western Region, 2+2 formation [43142 + 42257 + 44029 + 43141], between Old Oak Common and Bristol	43141 + 43142	[TGS]
July 24	HST Braking tests on Western Region, 2+2 formation [43142 + 42257 + 44029 + 43141], between Old Oak Common and Bristol	43141 + 43142	[TGS]
July 25	HST Braking tests on Western Region, 2+7 formation [43142 + 41021 + 42030 + 42031 + 42032 + 44010 + 42257 + 44029 + 43141], between Old Oak Common and Bristol	43141 + 43142	[TGS]
Aug 29	Bruff Road / Rail recovery vehicle - [B445WPO] Static tests at the RTC, then running tests at the Mickleover Test Line, including ride, braking and stress tests	-	-
Sep 17	HSFV1 RDB511023 ride test Derby - Cricklewood - Derby, Laden condition, tests up to 100 mile/h	Class 47	[TC1]
Sep 22	Mk III Coach M11025 Brake Tests (Ferodo & Becorit brake pads) Slip/Brake tests up to 100 mile/h between Crewe and Winsford on the down fast line	Class 86	[TC2]
Sep 26	Mk III Coach M11025 Brake Tests (Ferodo & Becorit brake pads) Slip/Brake tests up to 75 mile/h between Crewe and Winsford on the down slow line	Class 47	[TC2]
Sep 29	Mk III Coach M11025 Brake Tests (Ferodo & Becorit brake pads) Slip/Brake tests up to 100 mile/h between Crewe and Winsford on the down fast line	Class 86	[TC2]
Nov 4	HST Power Car (back to back) Brake Tests, Old Oak Common to Reading [43141 + 43142] Becorit brake pads	43141 + 43142	-
Nov 5	HST Power Car (back to back) Brake Tests, Old Oak Common to Reading [43141 + 43142] Becorit brake pads	43141 + 43142	-
Nov 6	HST Power Car (back to back) Brake Tests, Old Oak Common to Reading [43141 + 43142] Becorit brake pads	43141 + 43142	-
Nov 18	HST Braking tests on Western Region, 2+7 formation [43142 + 41129 + 41130 + 40436 + 42267 + 42268 + 42269 + 44032 + 43141], between Old Oak Common and Bristol	43141 + 43142	[TGS]
Nov 19	HST Braking tests on Western Region, 2+7 formation [43142 + 41129 + 41130 + 40436 + 42267 + 42268 + 42269 + 44032 + 43141], between Old Oak Common and Bristol	43141 + 43142	[TGS]

Diary 1985	Test Description	Loco	Test Car
Nov 20	HST Braking tests on Western Region, 2+7 formation [43019 + 41129 + 41130 + 40436 + 42267 + 42268 + 42269 + 44032 + 43024], between Old Oak Common and Bristol	43019 + 43024	[TGS]
Nov 21	HST Power Car (back to back) Brake Tests, Old Oak Common to Reading [43019 + 43024] Ferodo brake pads	43019 + 43024	[TGS]
Nov 27	HST Braking tests on Western Region, 2+7 formation [43192 + 41129 + 41130 + 40436 + 42267 + 42268 + 42269 + 44032 + 43141], between Old Oak Common and Bristol	43141 + 43192	[TGS]
Dec 12	Flatrol Wagon Ride Test DB900029 (laden) - short run on Midland Main Line	Class 47	[TC1]
Dec 16	Flatrol Wagon Ride Test DB900029 (tare) - short run on Midland Main Line	Class 47	[TC1]

5. TESTING DIARY - 1986

Diary 1986	Test Description	Loco	Test Car
Jan 14	45T Crane ADRC96717 riding and braking tests at Mickleover test line at speeds up to 45 mile/h	Class 47	[TC2]
Jan 15	45T Crane ADRC96717 handbrake tests at the RTC Derby	-	-
Jan 16	Plasser Long Welded Rail Train (LWRT) YXA Power Wagon (89010) tests up to 60 mile/h along the Mickleover Test line	Class 47	[TC2]
Jan 17	Plasser Long Welded Rail Train (LWRT) YXA Power Wagon (89010) handbrake tests at the RTC Derby	-	-
Jan 20	Railease Container Flat (RLS92648) Slip/Brake Tests up to 75 mile/h between Crewe and Winsford on the down slow line	Class 47	[TC2]
Jan 23	Instrumentation & strain gauge fitting on BSC Iron Ore Tippler Wagons at the RTC Derby	-	-
Jan 24	Instrumentation & strain gauge fitting on BSC Iron Ore Tippler Wagon (PTA 26100) at the RTC Derby	-	-
Jan 27	Autic 3-axle Car Carrier Wagon (RLS92050/1) Slip/Brake Tests - Crewe - Winsford	Class 47	[TC2]
Feb 1	BSC Iron Ore Tippler Wagon (PTA 26100) structural tests to Scunthorpe & Immingham, including loading and unloading of the wagons whilst under test conditions	47332	[TC1]
Feb 13	Autic 3-axle Car Carrier Wagon (RLS92050/1) Slip/Brake Tests - Crewe - Winsford	Class 47	[TC2]
Feb 21	Railease Container Flat (RLS92648) Slip/Brake Tests up to 75 mile/h between Crewe and Winsford on the down slow line	Class 47	[TC2]
Feb 24	102 tonne TEA Tank Wagon 84279 (Laden) - Slip/Brake Tests up to 60 mile/h between Crewe and Winsford on the down slow line	Class 47	[TC2]
Feb 25	102 tonne TEA Tank Wagon 84279 (Tare) - Slip/Brake Tests up to 60 mile/h between Crewe and Winsford on the down slow line	Class 47	[TC2]
Feb 26	102 tonne TEA Tank Wagon 84279 static brake tests and block force measurements	-	-
Feb 27	Bardon Hill 90 tonne Hopper PHA Wagon 17102 (Tare) Slip/Brake Tests up to 60 mile/h between Crewe and Winsford on the down slow line	Class 47	[TC2]
Mar 7	Flatrol Wagons Ride Test DB900030 (tare), DB900010 (laden) - Derby - Manton - Derby	Class 45	[TC1]
Mar 12	Flatrol Wagons Ride Test DB900030 (laden), DB900010 (tare) - Derby - Manton - Derby	45134	[TC1]

Diary 1986	Test Description	Loco	Test Car
Mar 13	Bardon Hill 90 tonne Hopper PHA Wagon 17102 (Laden) Slip/Brake Tests up to 60 mile/h between Crewe and Winsford on the down slow line	Class 47	[TC2]
Mar 20	102 tonne TEA Tank Wagon 84279 (Tare) - Slip/Brake Tests up to 60 mile/h between Crewe and Winsford on the down slow line	47196	[TC2]
Mar 23	Preparations for Stone train coupling and braking tests at Westbury	-	-
Mar 24	Stone Train braking and coupling tests - Westbury - Theale - Westbury (PHA Wagons)	56048 + 59039	-
Mar 25	Stone Train braking and coupling tests - Westbury - Theale - Westbury (trial of rear of train brake application device)	56048 + 59039	-
Mar 26	Stone Train braking and coupling tests - Westbury - Theale - Westbury (PTA Wagons)	56048 + 59039	-
Apr 10	Sheerness Steel PXA Bogie Scrap Metal Wagon (PR3150) fitted with Gloucester 3 Piece Bogies, ride test up to 60 mile/h on the Midland Main Line between Derby-Manton-Bedford-Derby	Class 47	[TC1]
Apr 14	Sheerness Steel PXA Bogie Scrap Metal Wagon (PR3150) Slip/brake test up to 60 mile/h between Crewe and Winsford on the down slow line	Class 47	[TC2]
Apr 21	Bardon Hill 90 tonne Hopper PHA Wagon 17102, in the Laden condition with modified brake forces, Slip/brake Tests up to 60 mile/h between Crewe and Winsford on the down slow line	Class 47	[TC2]
Apr 24	Plasser Excavator LDRP96505, Ride Test, Derby-Manton-Corby-Cricklewood-Derby	47538	[TC1]
Apr 28	Plasser Excavator LDRP96505 - Traction testing at the Mickleover Test Line, locomotive hauled from Derby to Mickleover	Class 31	[TC2]
May 8	Trailer Train (PXA) - TN95951-4, Gloucester GPS20 1.8m bogies, ride test up to 75 mile/h on the Midland Main Line between Derby-Manton-Cricklewood-Derby	47063	[TC1]
May 15	Trailer Train (PXA) - TN95951-4, Gloucester GPS20 1.8m bogies, Slip/brake test up to 75 mile/between Crewe and Winsford on the down slow line	Class 47	[TC2]
May 22	Bardon Hill 90 tonne Hopper PHA Wagon 17102 Slip/brake Tests with modified brake forces in the Tare condition, up to 60 mile/between Crewe and Winsford on the down slow line	47193	[TC2]

Diary 1986	Test Description	Loco	Test Car
June 16	Trailer Train (PXA) - TN95951-4, Gloucester GPS20 1.8m bogies, laden condition ride test up to 60 mile/h on the Midland Main Line between Derby-Manton-Bedford-Derby	45122	[TC1]
June 19	Trailer Train (PXA) - TN95951-4, Gloucester GPS20 1.8m bogies, Slip/brake laden condition tests up to 60 mile/h between Crewe and Winsford on the down slow line	Class 47	[TC2]
June 26	Tiphook 2-axle scrap wagon - 5756432-9, (Tare) ride test between Derby-Manton-Bedford-Derby	31127	[TC1]
June 30	Tiphook 2-axle scrap wagon - 5756432-9, (Laden) ride test between Derby-Corby-Derby	31152	[TC1]
July 1	Tiphook 2-axle scrap wagon - 5756432-9, handbrake tests at the RTC Derby	-	-
July 3	Tiphook 2-axle scrap wagon - 5756432-9, Wagon Laden Slip/brake Tests = Crewe-Winsford	Class 47	[TC2]
July 7	Transfesa 2-axle Ferry Van Ride Test, tare condition, Derby - Wellingborough - Derby	31127	[TC1]
July 8	Plasser Piling Vehicle (LDRP 96511) static torsional stiffness (ΔQ/Q) tests RTC	-	-
July 10	Sheerness Steel PXA Bogie Scrap Metal Wagon (PR3150) Slip/brake Test up to 60 mile/h between Crewe and Winsford on the down slow line	47190	[TC2]
July 14	Plasser Piling vehicle (LDRP 96511) ride test up to 60 mile/h on the midland Main line between Derby - Manton - Bedford - Derby	31418	[TC1]
July 18	Plasser piling vehicle (LDRP 96511) static handbrake tests and brake force tests - RTC	-	-
July 21	Plasser Piling vehicle (LDRP 96511) traction performance tests at Mickleover Test line	47049	[TC2]
July 25	Plasser Piling vehicle (LDRP 96511) Slip/brake Tests up to 60 mile/h between Crewe and Winsford on the down slow line.	47125	[TC2]
July 31	Sheerness Steel PXA Bogie Scrap Metal Wagon (PR3150) after modifications to the brake system load weigh, Slip/brake Tests up to 60 mile/h between Crewe and Winsford on the down slow line	Class 47	[TC2]
Aug 1 – 4	Preparation and instrument fitting for FY wagon stress tests - at the RTC Derby	-	[TC1]
Aug 5	Stress Tests on FY 102 tonne Aluminium Hopper Wagons up to 60 mile/h along the Midland Main Line and Great Western route between Derby - Brent - Westbury	31305	[TC1]

Diary 1986	Test Description	Loco	Test Car
Aug 6	Stress Tests on FY 102 tonne Hopper Wagons, Westbury to Merehead, loading tests at Merehead Quarry and running tests between Merehead and Westbury	47131	[TC1]
Aug 7	Stress Tests on FY 102 tonne Hopper Wagons returning from Westbury via the Western route to Reading, Brent then Midland Main line to Derby	47131	[TC1]
Aug 18	Powell Duffryn Steel Wagon PDUF3009 Slip/brake Test - Crewe-Winsford (return to RTC via Bescot), Class 31 transit + Class 86 electric for the testing up to 60 mile/h between Crewe and Winsford on the down slow line	Class 31 & Class 86	[TC2]
Aug 20	Class 37/7 braking performance static & dynamic testing of light locomotive at RTC Derby and between Derby and Manton Junction	37799	-
Aug 21	Ride test on Plasser 'Porpoise' Long Welded Rail Train (LWRT) 85 chute wagon DB979514, Y25 1.8m bogies, tests up to 60 mile/h between Eastleigh - Southampton - Salisbury, with portable equipment	Class 33	-
Sept 9	Powell Duffryn Steel Wagon PDUF3009 static brake characteristics and brake block force tests at the RTC Derby	-	-
Sept 11	Powell Duffryn Steel Wagon PDUF3009 Slip/brake Test - Crewe-Winsford (return to RTC via Bescot), Class 31 transit + Class 86 electric for the testing	Class 31 & Class 86	[TC2]
Sept 16	Flatrol Wagon [DB900010] static torsional stiffness ($\Delta Q/Q$) tests and suspension set-up checks RTC	-	-
Sep 25	Sheerness Steel PXA Bogie Scrap Metal Wagon (PR3150) fitted with Gloucester 3 Piece Bogies, ride test up to 60 mile/h on the Midland Main Line between Derby-Manton-Bedford-Derby	Class 47	[TC1]
Sept 29	Plasser 4 tonne GP Crane - LDRP96513 - traction tests at Mickleover Test line, locomotive hauled from Derby to Mickleover	Class 31	[TC2]
Oct 2	Sheerness Steel PXA Bogie Scrap Metal Wagon (PR3150) fitted with Gloucester 3 Piece Bogies, laden condition Slip/brake Tests up to 60 mile/h between Crewe and Winsford on the down slow line	31191 transit + 85007 testing	[TC2]
Oct 4	Resistance Testing HAA Wagons Train + Test Car 6 - Bristol – Taunton – Bristol	Class 47	[TC6]
Oct 5	Resistance Testing HAA Wagons Train + Test Car 6 - Bristol – Taunton – Bristol	Class 47	[TC6]

Diary 1986	Test Description	Loco	Test Car
Oct 7	Plasser PBI-84 (Pneumatic Ballast Injection Machine) DR73700, ride test under own power up to 45 mile/h with portable ride equipment between West Ealing-Reading-Newbury	-	-
Oct 9	Powell Duffryn Steel Wagon PDUF3009 Slip/brake Tests up to 60 mile/h between Crewe and Winsford on the down slow line	Class 47	[TC2]
Oct 11	Resistance Testing HAA Wagons Train + Test Car 6 - Bristol – Taunton	47284	[TC6]
Oct 12	Resistance Testing HAA Wagons Train + Test Car 6 - Bristol – Taunton	47284	[TC6]
Oct 14	GMC92580 -PFA Bogie Container Flat Wagon static tests RTC	-	-
Oct 19	GMC92580 -PFA Bogie Container Flat Wagon ride test up to 75 mile/h - Derby-Cricklewood-Derby	Class 47	[TC1]
Oct 27	Mentor Test run, Hitchin – Huntingdon all lines	-	[MEN]
Oct 29	Plasser 4 tonne Crane - LDRP96513 -wedge tests at the RTC Derby	-	-
Oct 30	Plasser 4 tonne Crane - LDRP96513 - wedge tests at the RTC Derby	-	-
Nov 5	Plasser 4 tonne Crane - LDRP96513 - ride test - Derby-Corby-Derby	31207	[TC1]
Nov 6	Mentor Test Run – Crewe – Glasgow – Crewe, main lines	-	[MEN]
Nov 7	Mentor Test Run – Crewe – Euston – Crewe, main lines	-	[MEN]
Nov 12	Plasser 4 tonne Crane - LDRP96513, ride retest, Derby - Bedford - Beeston - Derby. The train returned via Beeston in order to turn the Test Car ready for the next test with Freightliners.	Class 47	[TC1]
Nov 19	Freightliner wagons Ride Test, 601135 laden and 601239 tare condition, Derby - Beeston – Cricklewood –Derby. Testing up to 75 mile/h, Containers loaded onto 601135 at Beeston	47357	[TC1]
Nov 20	Freightliner wagons Ride Test, 601135 tare and 601239 laden Derby - Beeston – Cricklewood –Derby. Testing up to 75 mile/h, Containers swapped between wagons at Beeston	47357	[TC1]
Nov 21	Freightliner wagons, wedge tests / investigations at EDU workshop RTC	-	-
Nov 24	PFA Bogie Container Flat Wagon, RLS92627 bogie rotation tests at the RTC Derby	-	-
Nov 27	PFA Bogie Container Flat Wagon, RLS92627 Laden - Ride Test Derby-Cricklewood-Derby Testing up to 75 mile/h	Class 47	[TC1]

Diary 1986	Test Description	Loco	Test Car
Nov 28	PFA Bogie Container Flat Wagon, RLS92627 bogie rotation tests at the RTC Derby	-	-
Dec 2	Flatrol Wagons DB998014 and DB900122 Static Delta Q/Q Tests at the RTC Derby	-	-
Dec 11	Flatrol Wagons DB998014 and DB900122 Laden ride Test Derby-Corby-Derby	Class 47	[TC1]
Dec 18	Flatrol Wagons DB998014 and DB900122 Tare ride Test Derby-Corby-Derby	45132	[TC1]

6. TESTING DIARY - 1987

Diary1987	Test Description	Loco	Test Car
Jan 10	Tiphook Container Flat Wagons - GPS20 and Y25 1.8m bogie types - static torsional stiffness ($\Delta Q/Q$)tests at the RTC Derby	-	-
Jan 12	Tiphook Container Flat Wagons, GPS20 and Y25 bogie types - static bogie rotational resistance tests at the RTC Derby	-	-
Jan 14	Tiphook Container Flat Wagon, GPS20 and Y25 bogie types tare condition ride testing on the Midland Main line between Derby and Cricklewood up to 75 mile/h	Class 47	[TC1]
Jan 17	Tiphook Container Flat Wagon - GPS20 and Y25 bogie types laden condition ride testing on the Midland Main line between Derby and Cricklewood up to 75 mile/h	Class 47	[TC1]
Jan 22	Tiphook Container Flat Wagon, GPS20 and Y25 bogie types - slip test laden condition up to 75 mile/h between Crewe and Winsford on the down slow line	Class 47	[TC2]
Jan 24	Tiphook Container Flat Wagon, GPS20 and Y25 bogie types - slip test tare condition up to 75 mile/h between Crewe and Winsford on the down slow line	Class 47	[TC2]
Jan 28	Plasser Pneumatic Ballast Injection Machine - DR 73700, repeat ride tests, Derby - Kettering - Derby	Class 45	[TC1]
Jan 29	Plasser Pneumatic Ballast Injection Machine (PBI) - DR 73700, repeat ride tests, Derby - Finedon Road - Derby	45104	[TC1]
Feb 4	Class 154 DMU Acceptance and Performance Tests, Crush condition braking tests, Derby - Trent - Leicester - Derby	154001	-
Feb 5	Class 154 DMU Acceptance and Performance Tests, Crush condition, Lickey incline, starting performance tests, Derby - Birmingham NS - Bromsgrove	154001	-
Feb 6	Class 154 DMU Acceptance and Performance Tests, Crush condition, Buxton - Manchester - Hazel Grove	154001	-
Feb 11	Class 154 DMU Acceptance and Performance Tests, Crush condition, Lickey incline, starting performance tests, Derby - Birmingham NS - Bromsgrove & Longbridge	154001	-
Feb 13	Class 154 DMU Acceptance and Performance Tests, Crush condition braking and traction performance tests, Derby - Manton - Cricklewood - Derby	154001	-
Feb 20	Class 154 DMU Acceptance and Performance Tests, Fully seated condition, Lickey incline, starting performance tests, Derby - Birmingham NS - Northfield & Bromsgrove	154001	-

Diary1987	Test Description	Loco	Test Car
Feb 22 - 24	SPA / OCA Wagon bearing failure investigations, preparation works in the RTC Yard (Way & Works sidings), and instrumentation installation on wagon DC112082	-	[TC1]
Feb 25	SPA / OCA Wagon bearing failure investigations, works in the RTC Yard (Way & Works sidings), trial runs of impact operations.	08021	[TC1]
Feb 26	SPA / OCA Wagon bearing failure investigations, tests 1-3, FAG Bearings (new design), on wagon DC112082	08021	[TC1]
Mar 2	SPA / OCA Wagon bearing failure investigations, tests 4-7, FAG Bearings (standard design), on wagon DC112082	08021	[TC1]
Mar 4	SPA / OCA Wagon bearing failure investigations, tests 8-11, SKF bearings, on wagon DC112082	08021	[TC1]
Mar 9	SPA / OCA Wagon bearing failure investigations, tests 12-15, Timken bearings, on wagon DC112082	08021	[TC1]
Mar 12	SPA / OCA Wagon bearing failure investigations, tests 16-20, FAG Bearings (standard design with reduced suspension clearances), on wagon DC112082	08021	[TC1]
Mar 29	RH Roadstone 90 tonne Hopper Wagon RHR17301, GPS22.5 bogies, bogie rotational resistance test and static brake tests	-	-
Mar 30 - Apr 4	Class 37/7 modified braking performance static & dynamic testing of light locomotive at RTC Derby and between Derby and Crewe.	37799	-
May 8 - 10	LTF Bogie Tests YXA Wagon - (RDC921000), testing on the Midland Main Line route to Cricklewood, test running up to 75mile/h	Class 47	[TC1]
May 11 - 13	LTF Bogie Tests YXA Wagon - (RDC921000), testing on the Midland Main Line route to Cricklewood, test running up to 90mile/h	47609	[TC1]
May 14 - 15	LTF Bogie Tests YXA Wagon - (RDC921000), testing on the Midland Main Line route to Cricklewood, test running up to 90mile/h	Class 47	[TC1]
June 2	RH Roadstone PHA 90 tonne Hopper Wagon (RHR17301), GPS22.5 bogies, parking brake test	-	-
June 8	RH Roadstone 90 tonne Hopper Wagon RHR17301, GPS22.5 bogies, Slip/brake Tests up to 60 mile/h between Crewe and Winsford on the down slow line	Class 47	[TC2]
June 11	BSC Iron Ore Tippler Wagons (26671 + 26676), P6 worn wheel profiles - ride and stability test, Derby - Bedford - Derby, testing up to 60 mile/h	Class 47	[TC1]

Diary1987	Test Description	Loco	Test Car
June 18	DTS62-N (DR72201) Tests - Derby - Kettering - Derby, ride testing up to 60 mile/h	Class 31	[TC1]
June 23	Plasser 09-16 CSM, (DR73001) static torsional stiffness (ΔQ/Q) and bogie rotation tests at the RTC Derby	-	-
June 24	Class 150/1 DMU Track Recording Unit DB999600 + DB999601, static torsional stiffness (ΔQ/Q) test	-	-
June 29	Plasser 09-16 CSM, (DR73001) self-power tests at speeds up to 60 mile/h between Derby and Kettering	-	-
June 30	Plasser 09-16 CSM, (DR73001) self-power tests at speeds up to 60 mile/h between Derby and Manton	-	-
July 1	RH Roadstone PHA 90 tonne Hopper Wagon (RHR17301), GPS22.5 bogies, torsional stiffness (ΔQ/Q) and bogie rotational resistance test	-	-
July 6	RH Roadstone PHA 90 tonne Hopper Wagon (RHR17301), GPS22.5 bogies, up to 60 mile/h between Crewe - Winsford	47188	[TC2]
July 7	Plasser DTS62-N (DR72201), transit from Derby to Crewe hauled by locomotive then self-powered Braking Tests at speeds up to 60 mile/h between Crewe - Winsford	45041	[TC2]
July 12	Static Delta Q/Q tests on 2-axle Procor PFA Container flat wagon PR93209, at the RTC Derby	-	-
July 13 - 14	RH Roadstone PHA 90 tonne Hopper Wagon (RHR17301), GPS22.5 bogies, parking brake modifications and retest	-	-
July 20 - 23	BRE International Coach Tests - T4 Bogie	-	-
Aug 8	Static torsional stiffness (ΔQ/Q) tests TIPH93243 Tiphook Wagon at the RTC Derby	-	-
Aug 10	Tiphook Wagon TIPH93243 Ride Test up to 75 mile/h between Derby - Wellingborough (Finedon Rd) - Lawley St (Birmingham) - Derby. The wagon was loaded with test weight containers at Lawley Street Freightliner Terminal	37222	[TC1]
Aug 12	2-axle Procor PFA Container flat wagon PR93209 - repeat static tests (Delta Q/Q) - RTC	-	-
Aug 13	Tiphook Wagon TIPH93243 Slip/brake tests (Laden condition) up to 75 mile/h between Crewe and Winsford on the down slow line	47281	[TC2]
Aug 17	Tiphook Wagon TIPH93243 Ride Tests up to 75 mile/h on the Midland Main Line route between Derby - Manton - Cricklewood - Derby	31107	[TC1]

Diary1987	Test Description	Loco	Test Car
Sept 10	Tiphook Wagon TIPH93243 PFA Ride Tests up to 75 mile/h - Derby - Leicester - Derby	47281	[TC1] + [TC2]
Sept 17	Redland In-Line Tipper Wagon with GPS Bogies (REDA28100) static tests.	-	-
Oct 13 - 28	Preparations, static tests including bogie rotational resistance and torsional stiffness ($\Delta Q/Q$) tests and instrumentation installation on Mk III Coach 12140 fitted with SIG bogies	-	[TGS]
Oct 29	SIG Bogie Tests Mk III Coach M12140 instrument proving run out and back from the RTC Derby - Kettering - Derby, 100 mile/h maximum laden condition 12140 + 42317	43155 + 43102	[TGS]
Oct 30	Transit movement of the SIG bogie test train from the RTC Derby to Heaton (43102 + 44090 + 42317 + 12140 + 44101 + 43155) laden condition 12140 + 42317	43155 + 43102	[TGS]
Oct 31	SIG Bogie Tests Mk III Coach M12140 Darlington - York - Darlington (HST Power Cars 43102 + 44090 + 42317 + 12140 + 44101 + 43155) laden condition 12140 + 42317	43155 + 43102	[TGS]
Nov 1	SIG Bogie Tests Mk III Coach M12140 Darlington - York - Darlington (HST Power Cars 43102 + 44090 + 42317 + 12140 + 44101 + 43159) - World Diesel Speed Record, laden condition 12140 + 42317, 140 mile/h speed profile test, and stability check at 145 mile/h	43159 + 43102	[TGS]
Nov 2	SIG Bogie Tests Mk III Coach M12140 Newcastle - Bounds Green - Edinburgh - Newcastle (HST Power Cars 43102 + 44090 + 42317 + 12140 + 44101 + 43159), laden condition 12140 + 42317, 125 mile/h speed profile test	43159 + 43102	[TGS]
Nov 3	SIG Bogie Tests Mk III Coach M12140 Newcastle - Bonds Green - Edinburgh - Newcastle (HST Power Cars 43102 + 44090 + 42317 + 12140 + 44101 + 43159), laden condition 12140 + 42317, 140 mile/h speed profile test	43159 + 43102	[TGS]
Nov 4	SIG Bogie Tests Mk III Coach M12140 Darlington - York - Darlington (HST Power Cars 43102 + 44090 + 42317 + 12140 + 44101 + 43104), tare condition, 140 mile/h speed profile test, and stability check at 145 mile/h	43104 + 43102	[TGS]
Nov 5	SIG Bogie Tests Mk III Coach M12140 Newcastle - Bounds Green - Edinburgh - Newcastle (HST Power Cars 43102 + 44090 + 42317 + 12140 + 44101 + 43104), tare condition 12140 + 42317, 125 mile/h speed profile test	43104 + 43102	[TGS]
Nov 6	SIG Bogie Tests Mk III Coach M12140 Newcastle - Bounds Green - Newcastle (HST Power Cars 43102 + 44090 + 42317 + 12140 + 44101 + 43104), tare condition 12140 + 42317, 140 mile/h speed profile test	43104 + 43102	[TGS]

Diary1987	Test Description	Loco	Test Car
Nov 7	SIG Bogie Tests Mk III Coach M12140 Newcastle - Bounds Green - York (HST Power Cars 43102 + 44090 + 42317 + 12140 + 44101 + 43104), laden condition 12140 + 42317, 140 mile/h speed profile test. The test train then returned from York to Derby	43104 + 43102 43008	[TGS]
Nov 28	76 tonne Diesel Mechanical Breakdown Crane ADRC96200 - Slip/brake tests up to 60 mile/h between Crewe and Winsford on the down slow line	20097 & 47113	[TC2]
Dec 9	HAA Coal Hopper Wagon, 350453 tare condition - fitted with Ferodo AD7 Brake Pads, Slip/brake test on the down slow line between Crewe-Winsford, Loaded condition, up to 60 mile/h	47365	[TC2]

7. TESTING DIARY - 1988

Diary1988	Test Description	Loco	Test Car
Jan 13	76 tonne Diesel Mechanical Breakdown Crane, ADRC96200 static brake block force measurements at the RTC Derby	-	-
Jan 18	HAA Coal Hopper Wagon 350453fitted with Ferodo AD7 Brake Pads, 350453, Slip/brake test on the down slow line between Crewe-Winsford, Loaded condition, up to 60 mile/h	Class 47	[TC2]
Jan 21	HAA Coal Hopper Wagon 350453fitted with Ferodo AD7 Brake Pads,350453, Slip/brake test on the down slow line between Crewe-Winsford, Loaded condition, up to 60 mile/h	Class 47	[TC2]
Jan 25	Ride Comfort portable test instrumentation proving test Derby-Matlock (Class 150 Service Train) in preparation for the Class 455 project	-	-
Jan 27	Ride Comfort portable test instrumentation proving test Derby-Matlock (Class 150 Service Train) in preparation for the Class 455 project	-	-
Jan 28	Coles Rail Crane DRF81312 derailment investigation ride test, Derby - Corby - Derby	Class 47	[TC1]
Jan 29	Class 455/7 (Unit 5701) Ride Comfort Testing, Orpington to Ashford with portable test instrumentation, at speeds up to 75 mile/h	5701	-
Feb 9	Class 58 Locomotive windscreen demister tests Toton – Cricklewood – Toton	58023	-
Feb 12	Class 87 Windscreen demister tests Longsight - Manchester - London - Longsight	87019	-
Feb 29	Plasser DTS-62N (DR72202) Ride Test at speeds up to 60 mile/h between Derby - Manton - Derby [Self Power]	-	-
Mar 29	YBA Sturgeon & YPA Tench (Modified Sturgeon) Wagons (Plate Frame 1676mm bogies) - Ride / Axlebox Temperature Tests - Derby - Kettering - Derby	Class 31	[TC1]
Apr 6	YBA Sturgeon & YPA Tench (Modified Sturgeon) Wagons (Plate Frame 1676mm bogies)- Ride / Axlebox Temperature Tests - Derby - Kettering – Derby	31415	[TC1]
Apr 12	Class 73 Locomotive Test (Ride) Stewarts Lane - Orpington - Ashford (portable test instrumentation). The test train formation included a Class 423, 4VEP EMU unit	73210 + 4VEP EMU	-
Apr 14	YBA Sturgeon & YPA Tench (Modified Sturgeon), (Plate Frame 1676mm bogies) - Tyre Profile Measurements	-	-
May 4	Cardiff Cathays visit for assessment of new Ultrasonic Test Unit power cars - 977391 (ex-51433) + 977392 (ex-53167)	-	-
May 5	YXA RDC921000 Low Track Force (LTF) Bogie wagon static tests RTC Derby	-	-

Diary1988	Test Description	Loco	Test Car
May 10	YXA RDC921000 Low Track Force (LTF) Bogie Wagon Slip/brake Test tests up to 75 mile/h between Crewe and Winsford on the down slow line	Class 47	[TC2]
May 12	Plasser Heavy-Duty Self Propelled Twin Jib Crane (DRC7823x), hydraulic pressures measurements at Carlisle Upperby Yard [Static Test with instrumentation in the back of a Montego estate car]	-	-
May 17	YXA RDC921000 Low Track Force (LTF) Bogie Wagon Slip/brake Test tests up to 75 mile/h between Crewe and Winsford on the down slow line	Class 47	[TC2]
May 24	Preparation for Lucas Girling WSP System braking Tests on Mk III coach 17174	-	-
May 25	Preparation for Lucas Girling WSP System braking Tests on Mk III coach 17174	-	-
June 6 - 8	Lucas Girling WSP (Wheel Slide Prevention) System braking Tests on Mk III coach 17174 up to 75 mile/h between Crewe and Winsford on the down slow line. (TC2 + 17174 + RDB975428 + TC1)	Class 47	[TC2]
June 9 - 10	Lucas Girling WSP (Wheel Slide Prevention) System braking Tests on Mk III coach 17174 up to 75 mile/h between Crewe and Winsford on the down slow line. (TC2 + 17174 + RDB975428 + TC1)	47545	[TC2]
June 13 - 17	Lucas Girling WSP (Wheel Slide Prevention) System braking Tests on Mk III coach 17174 up to 75 mile/h between Crewe and Winsford on the down slow line. (TC2 + 17174 + RDB975428 + TC1)	Class 47	[TC2]
June 20	Lucas Girling WSP (Wheel Slide Prevention) System braking Tests on Mk III coach 17174 up to 75 mile/h between Crewe and Winsford on the down slow line. (TC2 + 17174 + RDB975428 + TC1)	47545	[TC2]
June 21 - 23	Lucas Girling WSP (Wheel Slide Prevention) System braking Tests on Mk III coach 17174 Crewe- Winsford. (TC2 + 17174 + RDB975428 + TC1)	Class 47	[TC2]
June 24	Lucas Girling WSP (Wheel Slide Prevention) System braking Tests on Mk III coach 17174 up to 75 mile/h between Crewe and Winsford on the down slow line. (TC2 + 17174 + RDB975428 + TC1)	47545	[TC2]
July 25	SIG Bogie prototype tests with Mk III Coach 12140 Derby - Leicester – Derby 47561 + RDB975547 + 10706 + 12140 + 17174 + TC6	47561	[TC6]
July 26	SIG Bogie prototype tests with Mk III Coach 12140 Derby - Crewe - Carlisle - Crewe, testing up to 110 mile/h. 87023 + RDB975547 + 10706 + 12140 + 17174 + TC6	47561 & 87023	[TC6]

Diary1988	Test Description	Loco	Test Car
July 27	SIG Bogie prototype tests with Mk III Coach 12140 Crewe - Carlisle - Crewe - Derby, testing up to 100 mile/h. 81007 + RDB975547 + 10706 + 12140 + 17174 + TC6	81007 & Class 47	[TC6]
July 28	SIG Bogie prototype tests with Mk III Coach 12140 Derby - Crewe - Willesden - Crewe - Derby, testing up to 100 mile/h. 47512 + RDB975547 + 10706 + 12140 + 17174 + TC6	47512	[TC6]
Aug 2	SIG Bogie prototype tests with Mk III Coach 12140 Derby - Crewe - Carlisle - Crewe , testing up to 110 mile/h, Class 87 + RDB975547 + 10706 + 12140 + 17174 + TC6	Class 87	[TC6]
Aug 3	SIG Bogie prototype tests with Mk III Coach 12140 Crewe - Carlisle - Crewe ETD, testing up to 100 mile/h, Class 47 + RDB975547 + 10706 + 12140 + 17174 + TC6	Class 47	[TC6]
Aug 11	Powell Duffryn small-wheeled bogie wagon PDUF95301 (PXA), Ride test - Midland Main Line, experimental trial vehicle conversion from a BGV steel carrying wagon W160030 which was originally built as a Warflat wagon in 1943, the conversion involved the fitting of the prototype small-wheeled Gloucester bogies	Class 47	[TC1]
Aug 15	Powell Duffryn small-wheeled bogie wagon PDUF95301 (PXA), Slip/brake test up to 75 mile/h between Crewe and Winsford on the down slow line	Class 47	[TC2]
Aug 18	Powell Duffryn small-wheeled bogie wagon PDUF95301 (PXA), Slip/brake test - Crewe-Winsford	Class 47	[TC2]
Sept 16	Clarke Chapman (Cowans Sheldon) 12 tonne General Purpose Crane Ride Test, Derby - Bedford - Derby, test speed limited to 50 mile/h due to riding properties of the GP Crane	Class 47	[TC1]
Sept 29	Plasser Piling Machine Test - Derby - Beeston - Bedford - Derby, tests up to 60 mile/h.	Class 47	[TC1]
Oct 3	ARC (PHA) Wagon SRW17901 Slip/brake Test up to 75 mile/h between Crewe and Winsford on the down slow line	Class 47	[TC2]
Oct 4	ARC (PHA) Wagon SRW17901 Slip/brake Test up to 75 mile/h between Crewe and Winsford on the down slow line	Class 47	[TC2]
Oct 10	ARC (PHA) Wagon SRW17901 Ride & Axle Stress Tests - Derby - Cricklewood	Class 47	[TC1]
Oct 17	ARC (PHA) Wagon SRW17901 Ride & Axle Stress Tests - Derby - Cricklewood	Class 47	[TC1]
Oct 18	ARC (PHA) Wagon SRW17901 Ride & Axle Stress Tests - Derby - Bedford	Class 47	[TC1]
Oct 19	ARC (PHA) Wagon SRW17901 Slip/brake Test up to 75 mile/h between Crewe and Winsford on the down slow line.	Class 47	[TC2]

Diary1988	Test Description	Loco	Test Car
Oct 20	ARC (PHA) Wagon SRW17901 Slip/brake Test up to 75 mile/h between Crewe and Winsford on the down slow line.	Class 47	[TC2]
Nov 1	Handbrake performance test on Class 08 shunting locomotive at the RTC Derby	08842	-
Nov 8	ARC Wagon SRW17901 Ride Test up to 75 mile/h - Derby - Cricklewood - Derby	Class 47	[TC1]
Nov 14	O&K PHA 102 tonne Box Wagons (DB Type 25-100 2m bogies) OK3268/OK3269 Ride Test Derby - Bedford - Derby (Tare condition)	Class 47	[TC1]
Nov 19	O&K PHA 102 tonne Box Wagons (DB Type 25-100 2m bogies) OK3268/OK3269 Ride Test Derby - Westbury (Laden condition)	Class 47	[TC1]
Nov 20	O&K PHA 102 tonne Box Wagons (DB Type 25-100 2m bogies) OK3268/OK3269 Ride Test Westbury - Derby	Class 47	[TC1]
Nov 22	O&K PHA 102 tonne Box Wagons (DB Type 25-100 2m bogies) OK3268/OK3269, static braking and parking brake test	-	-
Nov 24	O&K PHA 102 tonne Box Wagons (DB Type 25-100 2m bogies) OK3268/OK3269 Laden Slip/brake Test up to 60 mile/h between Crewe and Winsford on the down slow line. The transit back to Derby was delayed due to a Brake Pipe fracture on Test Car 2 at Tutbury	Class 47	[TC2]
Nov 29	O&K PHA 102 tonne Box Wagons (DB Type 25-100 2m bogies) OK3268/OK3269 Tare Slip/brake Test up to 60 mile/h between Crewe and Winsford on the down slow line	Class 47	[TC2]
Dec 5	Tiphook (TIPH 96500) PXA Swingbed Inter-Modal Wagon Ride Test - Derby-Bedford-Derby	Class 47	[TC1]
Dec 10	Tiphook (TIPH 96500) PXA Swingbed Inter-Modal Wagon Ride Test - Derby-Cricklewood-Derby	Class 47	[TC1]
Dec 15	Tiphook (TIPH 96500) PXA Swingbed Inter-Modal Wagon static brake block load tests at the RTC Derby	-	-
Dec 17	Tiphook (TIPH 96500) PXA Swingbed Inter-Modal Wagon Slip/brake Test up to 75 mile/h between Crewe and Winsford on the down slow line	Class 47	[TC2]

8. TESTING DIARY - 1989

Diary1989	Test Description	Loco	Test Car
Jan 9	Tiphook (TIPH 96500) PXA Swingbed Inter-Modal Wagon Slip/brake Test up to 75 mile/h between Crewe and Winsford on the down slow line	Class 47	[TC2]
Jan 14	Foster Yeoman PGA 50 tonne Hopper Wagon (BSC Suspension) - PR14055, and Foster Yeoman PHA 102 tonne (GPS-25 Suspension) - PR17802, Laden condition ride test, Derby - Bedford - Derby. (Data collection as part of 25 tonne axleload wagon ride investigations)	Class 47	[TC1]
Jan 23	Start of Setup On-board Analysis system on Test Car 10ready for Mk IV Prototype Tests	-	[TC10]
Jan 27	Static torsional stiffness ($\Delta Q/Q$) test on Mk IV Coach 12201	-	-
Jan 28	Tarmac Quarry Products PGA 50 tonne Hopper Wagon (Gloucester Pedestal Suspension) - PR14243, and BSC 102 tonne Tippler Box Wagon PR26833, Laden condition ride test, Derby - Bedford - Derby. (Data collection as part of 25 tonne axleload wagon ride investigations)	Class 97	[TC1]
Feb 8	Instrument proving test run with Mk IV Coach 12201 and Test Car 10 - Derby - Bedford - Derby, max speed 100 mile/h	97545	[TC10]
Feb 9	Instrument proving and resistance to derailment data collection test run with Mk IV Coach 12201 - Derby - Bedford – Derby, max speed 100 mile/h	97545	[TC10]
Feb 10	HST Power Cars 43116 & 43051 arrive at RTC, Cables connected preparations for Mk IV 12201 Coach Tests	43116 + 43051	[TC10]
Feb 11	Transit Derby - Grantham, Tests Grantham - Connington - Grantham, stability tests including tests with yaw dampers removed, then Transit to KX and Bounds Green [43116+44019+12201+TC10+43051]	43116 + 43051	[TC10]
Feb 13	Mk IV Test runs Bounds Green - Kings X - Edinburgh - Heaton [43116+44019+12201+TC10+43051], testing up to 125 mile/h	43116 + 43051	[TC10]
Feb 14	Mk IV Test runs Heaton - Kings X - Edinburgh - Heaton [43116+44019+12201+TC10+43051], testing up to 125 mile/h	43116 + 43051	[TC10]
Feb 15	Installation of bogie cross-member to simulate BT41-C bogie derivative that will be fitted to the Mk IV DVT vehicles	-	-
Feb 16	Mk IV Test runs Heaton - Kings X - Edinburgh - Heaton [43116+44019+12201+TC10+43051], testing up to 125 mile/h	43116 + 43051	[TC10]

Diary1989	Test Description	Loco	Test Car
Feb 18	Mk IV Test runs Heaton - Kings X - Edinburgh - Heaton - Derby [43116+44019+12201+TC10+43051], including testing with air suspension deflated	43116 + 43051	[TC10]
Feb 25	SNCF Multifret Wagon Ride Test (Laden), Derby-Cricklewood-Derby	97545	[TC1]
Feb 27	SNCF Multifret Wagon Slip/brake Test (Laden) up to 75 mile/h between Crewe and Winsford on the down slow line	Class 97	[TC2]
Mar 13	O&K PHA 102 tonne Box Wagons (DB Type 25-100 2m bogies) OK3271/OK3272 static torsional stiffness (ΔQ/Q) and Bogie Rotation tests - RTC, following modifications including fitting UIC sidebearers	-	-
Mar 17	O&K PHA 102 tonne Box Wagons (DB Type 25-100 2m bogies) OK3271/OK3272 (Tare) Ride Test - Derby-Bedford-Bardon Hill (to Load the Wagons) - Derby	97545	[TC1]
Mar 18	O&K PHA 102 tonne Box Wagons (DB Type 25-100 2m bogies) OK3271/OK3272 (Laden) Ride Test - Derby-Bedford-Derby	97545	[TC1]
Mar 20	Tiphook PIA Hopper Wagon (690-5-033-9P) Y25 bogies, static torsional stiffness (ΔQ/Q) and Bogie Rotation Tests [RTC], (Data collection as part of 25 tonne axleload wagon ride investigations)	-	-
Mar 30	Mk IV Coach static Bogie Rotation Tests at the RTC Derby, with various different rate yaw dampers fitted	-	-
Apr 7	O&K PHA 102 tonne Box Wagons (DB Type 25-100 2m bogies) OK3271/OK3272 - static torsional stiffness (ΔQ/Q)– RTC	-	-
Apr 8	Tiphook PIA Hopper Wagon (690-5-033-9P) Y25 bogies, ride test Laden, Derby - Cricklewood - Derby	Class 97	[TC1]
Apr 17	Tope ZCV (converted HTV) 2-axle engineers wagon DB970047, Laden, converted from HTV hopper wagons, Ride Test, Derby - Kettering - Derby, testing up to 50 mile/h	Class 97	[TC1]
Apr 19	Tope ZCV (converted HTV) 2-axle engineers wagon DB970047, Tare, Ride Test, Derby - Kettering - Derby, testing up to 50 mile/h	Class 97	[TC1]
May 4	YXA RDC921000 Low Track Force (LTF) Bogie Wagon Test - Derby - Crewe - Bletchley - Crewe - Derby, test running up to 100 mile/h, Class 86 locomotive used for the WCML section of the test run	Class 97 & Class 86	[TC1]
May 9	72 tonne GLW Phosphorous Tank Wagon, DB Leaf Sprung 1.8m bogies, static torsional stiffness (ΔQ/Q) and Bogie Rotation Tests [RTC]	-	-

Diary1989	Test Description	Loco	Test Car
May 15	Plasser EPV-360 Excavator (LDRP96504) &GPC-38, 4 tonne Crane (LDRP96512) Ride re-tests at speeds up to 60 mile/h between Derby - Corby – Derby	Class 97	[TC1]
May 18	72 tonne Phosphorous Tank Wagon, DB Leaf Sprung 1.8m bogies, Ride Tests - Derby - Kettering - Derby	Class 97	[TC1]
June 7-8	Clam (ZCV) converted HTV Hopper Wagon DB973151 static torsional stiffness (ΔQ/Q) testing – RTC	-	-
June 19	O&K 102 tonne Hopper Wagons (DB Type 25-100 2m bogies) OK19300/OK19301 (Tare) Ride Test - Derby-Bedford-Bardon Hill (to Load the Wagons) - Derby	31416	[TC1]
June 20	O&K 102 tonne Hopper Wagons (DB Type 25-100 2m bogies) OK19300/OK19301 (Laden) Ride Test - Derby - Bedford - Derby	20173 + 20208	[TC1]
June 22-28	Preparation work at the RTC Derby for Mk IV Coach 12400 testing following modifications. [TC10 + 12200 + 12400 + 12300 + 12201]	-	[TC10]
June 29	Instrument proving test run with Mk IV Coach - Derby - Bedford - Derby, to a maximum speed f 100 mile/h. [TC10 + 12200 + 12400 + 12300 + 12201]	Class 47	[TC10]
June 30	Transit to Grantham then test proving runs with a Class 91 locomotive on the ECML, Peterborough to Grantham, including brake tests [TC10 + 12200 + 12400 + 12300 + 12201], up to 125 mile/h	Class 91 47973	[TC10]
July 1	Test proving runs with a Class 91 locomotive on the ECML, Peterborough to Grantham MK IV Coach Tests and return to Derby [TC10 + 12200 + 12400 + 12300 + 12201]	Class 91 47973	[TC10]
July 3	Transit Derby to Neville Hill Class 47, MK IV Coach Tests - 2 x HST Power Cars - Neville Hill - Doncaster - Edinburgh - Heaton [TC10 + 12200 + 12400 + 12300 + 12201]	Class 47, Class 43 & Class 43	[TC10]
July 4	MK IV Coach Tests - 2 x HST Power Cars - Heaton - Retford - Edinburgh - Heaton [TC10 + 12200 + 12400 + 12300 + 12201]	Class 43 + Class 43	[TC10]
July 6	Transit Mk IV Test train Heaton to Bounds Green [TC10 + 12200 + 12400 + 12300 + 12201]	Class 47	[TC10]
July 7	Mk IV Coach Tests Kings X - Doncaster - Kings X - Doncaster - Bounds Green, including brake tests [TC10 + 12200 + 12400 + 12300 + 12201]	Class 91	[TC10]

Diary1989	Test Description	Loco	Test Car
July 8	Mk IV Coach Tests Kings X - Doncaster - Kings X - Doncaster – Derby [TC10 + 12200 + 12400 + 12300 + 12201]	Class 91 & Class 47	[TC10]
July 19	8-axle Nuclear Flask Wagon, bogie rotation test, EDU	-	-
July 20	Instrument fitting in a 20 tonne CAR type, 2-axle Brake Van at Peak Forest yard in preparation for braking tests with stone train	-	-
July 23	Peak Forest - Buxton stone train braking tests with 2 x Class 37 locomotives (runaway investigation)	2 x Class 37	Brake Van
July 24	Peak Forest -Tinsley - Leeds - Tinsley - Peak Forest stone train braking tests with 2 x Class 37 locomotives (runaway investigation)	2 x Class 37	Brake Van
Aug 8	PD Lowliners PFA [PDUF95091/2] Ride Test up to 75 mile/h - Derby - Cricklewood - Lawley St (Load Containers) – Derby	47971	[TC1]
Aug 9	PD Lowliners PFA [PDUF95091/2] - Derby -Lawley St (Load Containers) – Derby	47481	[TC1]
Aug 10	PD Lowliners PFA [PDUF95091/2] Ride Test - test run cancelled before departure from the RTC due to locomotive failure	47973 failed	[TC1]
Aug 11	PD Lowliners PFA [PDUF95091/2] Ride Test - test run cancelled shortly after departure at Spondon due to locomotive failure	47971 failed	[TC1]
Aug 14	PD Lowliners PFA [PDUF95091/2] Ride Test up to 75 mile/h - Derby - Kentish Town - Derby, problems with access at Cricklewood meant that the run-round was unusually effected at Kentish Town	47973	[TC1]
Aug 16	PD Lowliners PFA [PDUF95091/2] Slip/brake Test up to 75 mile/h. transit from Derby to Crewe, testing on the down slow line to Winsford, then return transit via Lawley St to unload the test weight containers before returning to Derby	Class 47	[TC2]
Aug 18	Mk III DVT Slip/brake Tests at speeds up to 100 mile/h on the down fast line between Crewe and Winsford using a Class 86 electric locomotive	Class 47 & Class 86	[TC2]
Sept 6	Permaquip Multicar 3.5t Road/Rail vehicle test at Annesley Colliery line (Permaquip) - (F511LRR)	-	-
Sept 7	PD Lowliners [PDUF95091/2] Slip/brake Test - Derby - Crewe-Winsford - Derby	47972	[TC2]

Diary1989	Test Description	Loco	Test Car
Sept 12	Static torsional stiffness (ΔQ/Q) at RTC on Bruff Landrover (D695AVT) and Permaquip Multicar (F511LRR) Road Rail vehicles	-	-
Sept 13	Testing of Bruff Landrover (D695AVT) at Mickleover Test Line and Toton Depot Riding/Braking	-	-
Sept 14	Testing of Permaquip 3.5t Multicar (F511LRR) at Mickleover Test Line and Toton Depot Riding/Braking	-	-
Oct 17	O&K 102 tonne PHA Box Wagons (DB Type 25-100 2m bogies) OK3271/OK3272 ride tests with Brunninghaus 4+2 springs Tare - Derby - Bedford - Derby	47971	[TC1]
Oct 19	O&K 102 tonne PHA Box Wagons (DB Type 25-100 2m bogies) OK3271/OK3272 ride tests with Brunninghaus 4+2 springs Laden - Derby - Bedford - Derby. Buffer damage on TC1 before departure, replacement buffer taken from DMU 53119 and fitted before departure	47971	[TC1]
Oct 20	Mk IV Coaches arrived at RTC in preparation for the next stage of Mk IV coach acceptance tests, vehicles 12300 & 11200.	-	-
Oct 26	O&K 102 tonne PHA Box Wagons (DB Type 25-100 2m bogies) OK3271-OK3272static torsional stiffness (ΔQ/Q) and Bogie Rotation tests – RTC	-	-
Nov 1 - 15	Instrumentation fitting to Mk IV coaches 11200 & 12300 at the RTC Derby in preparations for ECML tests	-	-
Nov 17	Test instrument proving run - Mk IV Test Train, Derby - Bedford -Derby, formation 12204 + 12405 + 12300 + 11200 + 10305 + 82204	Class 47	[FO]
Nov 20	Mk IV Test Train transit move Derby - Bounds Green, 12204 + 12405 + 12300 + 11200 + 10305 + 82204	47973	[FO]
Nov 21	Mk IV Testing (Day 1) Bounds Green - Stevenage - Grantham (x4 runs) - Bounds Green, 91008 + 12204 + 12405 + 12300 + 11200 + 10305 + 82204, testing up to 140 mile/h	91008	[FO]
Nov 22	Mk IV Testing (Day 2) Bounds Green - Stevenage - Grantham (x4 runs) - Bounds Green, 91008 + 12204 + 12405 + 12300 + 11200 + 10305 + 82204, testing up to 140 mile/h	91008	[FO]
Nov 24	Mk IV Testing (Day 3) Bounds Green - Stevenage - Grantham (x4 runs) - Bounds Green, 91008 + 12204 + 12405 + 12300 + 11200 + 10305 + 82204, testing up to 140 mile/h	91008	[FO]
Nov 27	Mk IV Testing (Day 4) Bounds Green - Stevenage - Grantham (x4 runs) - Bounds Green, 91008 + 12204 + 12405 + 12300 + 11200 + 10305 + 82204, testing up to 140 mile/h	91008	[FO]

Diary1989	Test Description	Loco	Test Car
Nov 28	Mk IV Testing (Day 5) Bounds Green - Stevenage - Grantham (x4 runs) - Bounds Green (DVT Horn Failure, run 4 aborted), 91008 + 12204 + 12405 + 12300 + 11200 + 10305 + 82204, testing up to 140 mile/h	91008	[FO]
Nov 29	Mk IV Testing (Day 6) Bounds Green - Stevenage - Grantham (x4 runs) - Bounds Green locomotive (DVT Radio Failure), 91008 + 12204 + 12405 + 12300 + 11200 + 10305 + 82204, testing up to 140 mile/h	91008	[FO]
Nov 30	Mk IV Testing (Day 7) Bounds Green - Stevenage - Grantham (x4 runs) - Bounds Green, 91008 + 12204 + 12405 + 12300 + 11200 + 10305 + 82204, testing up to 140 mile/h	91008	[FO]
Dec 1	Mk IV Testing (Day 8) Bounds Green - Stevenage - Grantham (x4 runs) - Bounds Green, 91008 + 12204 + 12405 + 12300 + 11200 + 10305 + 82204, testing up to 140 mile/h	91008	[FO]
Dec 2	Mk IV Testing (Day 9) Bounds Green - Stevenage - Grantham (x2) - Bounds Green, 91008 + 12204 + 12405 + 12300 + 11200 + 10305 + 82204, testing up to 140 mile/h	91008	[FO]
Dec 4	Mk IV Testing (Day 10) Bounds Green - Stevenage - Grantham (x4) - Bounds Green, 91008 + 12204 + 12405 + 12300 + 11200 + 10305 + 82204, testing up to 140 mile/h	91008	[FO]
Dec 5	Mk IV Testing (Day 11) Bounds Green - Stevenage - Grantham (x4) - Bounds Green, 91008 + 12204 + 12405 + 12300 + 11200 + 10305 + 82204, testing up to 140 mile/h	91008	[FO]
Dec 6	Mk IV Testing (Day 12) Bounds Green - Stevenage - Grantham (x4) - Bounds Green, 91008 + 12204 + 12405 + 12300 + 11200 + 10305 + 82204, testing up to 140 mile/h	91008	[FO]
Dec 8	Mk IV Testing (Day 13) Bounds Green - Stevenage - Grantham (x2) - Bounds Green, 91008 + 12204 + 12405 + 12300 + 11200 + 10305 + 82204, testing up to 140 mile/h	91008	[FO]
Dec 9	Mk IV Testing (Day 14) Bounds Green - Stevenage - Peterborough, 91008 + 12204 + 12405 + 12300 + 11200 + 10305 + 82204, testing up to 125 mile/h, then transit back to Derby RTC	91008 Class 47	[FO]
Dec 11	Mk IV Testing Derby-Crewe-Carlisle-Rugby-Crewe-Derby	47971 & 90003	[FO]
Dec 18	Multifret Wagon ride Test (Tare) Derby-Cricklewood-Derby	47971	[TC1]

9. TESTING DIARY - 1990

Diary1990	Test Description	Loco	Test Car
Jan 3	Multifret Wagon Laden Ride Test - Derby-Cricklewood-Derby	47971	[TC1]
Jan 10	Speno RPS-32 Rail Grinder (DR79226) - static bogie rotation tests RTC	-	-
Jan 22	LTF and FBT6 bogie BAA wagons (RDC921000 + BAA900283) - Axle Strain Comparison Tests - Day 1, Derby - Peterborough - Cambridge - Stratford - Reading	Class 47	[TC4]
Jan 24	LTF and FBT6 bogie BAA wagons (RDC921000 + BAA900283) - Axle Strain Comparison Test run - Day 2, Reading - Southampton - Salisbury - Westbury - Reading - Didcot - Birmingham - Burton-on-Trent - Derby	Class 47	[TC4]
Jan 25	LTF and FBT6 bogie BAA wagons (RDC921000 + BAA900283) - Axle Strain Comparison Test run - Day 3, Derby - Crewe - Carnforth - Barrow - Carnforth - Leeds - Derby	Class 47	[TC4]
Jan 27	Static Body Sway Tests Mk III Coach 12300 Laden - RTC	-	-
Jan 28	Static Body Sway Tests Mk III Coach 12300 Laden - RTC	-	-
Jan 30	UIC 2-axle Tiger Tank Wagon (23 70 749 0 418-4), static torsional stiffness ($\Delta Q/Q$) test	-	-
Feb 5	UIC 2-axle Tiger Tank Wagon (23 70 749 0 418-4) Ride Test - Tare condition, Derby - Kettering - Derby; poor riding properties lead to a shortened test run, max speed 40 mile/h	Class 47	[TC1]
Feb 6	Speno RPS-32 Rail Grinder (DR79226) Static Bogie Rotation Tests RTC	-	-
Feb 8	UIC 2-axle Tiger Tank Wagon Ride Test - Laden - Derby - Kettering - Derby, test running up to 60 mile/h	Class 47	[TC1]
Feb 14	Speno RPS-32 Rail Grinder (DR79226) Riding Test - Derby - Oakham - Derby [Self Power], tests up to 60 mile/h	-	-
Feb 15	Speno RPS-32 Rail Grinder (DR79226) Braking Test - Derby - Leicester - Trent - Derby [Self Power] , tests up to 60 mile/h	-	-
Feb 26	HST Power Car + TGS coach 44027 Braking Tests Derby-York. Test cancelled soon after departure (reason not recorded)	-	-
Mar 1	British Rail Research Lab 5 (977469) Track Recording Coach Bogie Rotation tests (disc brakes fitted) - static test at the RTC Derby	-	[LAB5]
Mar 5	Start of instrument fitting on new Mk IV coaches 12478 and 12300 at RTC Derby	-	-
Mar 20	MK IV DVT + 3 Mk IV Coaches (10305 + 12400 + 12478) arrive at the RTC static weighing of wheel weights	47972	-

Diary1990	Test Description	Loco	Test Car
Mar 23	Instrument proving test run Mk IV coach tests Derby - Crewe – Carlisle, TOE + 12478 + 12400 + 12300 + 10305 + 82220 (1)	47972	[SV]
Mar 30	Instrument proving test run Mk IV coach tests Derby - Bedford – Derby, TOE + 12478 + 12400 + 12300 + 10305 + 82220 (2)	-	[SV]
Apr 2	Mk IV Ride Tests - Derby - Crewe - Carlisle - Crewe - Derby, TOE + 12478 + 12400 + 12300 + 10305 + 82220 (3)	31970 & 87018	[SV]
Apr 3	Mk IV Ride Tests - Derby - Crewe - Carlisle - Crewe – Derby, TOE + 12478 + 12400 + 12300 + 10305 + 82220 (4)	47972 & 87001	[SV]
Apr 4	Mk IV Ride Tests - Derby - Crewe - Preston - Crewe – Derby, TOE + 12478 + 12400 + 12300 + 10305 + 82220 (5)	47972	[SV]
Apr 5	Mk IV Ride Tests - Derby - Crewe - Carlisle - Crewe – Derby, TOE + 12478 + 12400 + 12300 + 10305 + 82220 (6)	47972 & 90034	[SV]
Apr 6	Mk IV Ride Tests - Derby - Crewe - Carlisle - Crewe – Derby, TOE + 12478 + 12400 + 12300 + 11200 + 10305 + 82220 (7)	47972 & 86426	[SV]
Apr 7	Static torsional stiffness ($\Delta Q/Q$) on Mk IV Coach 12400 after Anti-Roll Bar modifications – RTC	-	-
Apr 9	Mk IV Ride Tests - Derby - Crewe - Carlisle - Crewe – Derby, TOE + 12478 + 12400 + 12300 + 10305 + 82220 (8)	47971 & 87013	[SV]
Apr 10	Mk IV Ride Tests - Derby - Crewe - Carlisle - Crewe - Derby, TOE + 12478 + 12400 + 12300 + 10305 + 82220 (9)	47972 & Class 87	[SV]
Apr 23	Mk IV Ride Tests - Derby - Crewe - Carlisle - Crewe – Derby, TOE + 12478 + 12400 + 12300 + 10305 + 82220 (10)	47971 & 90024	[SV]
Apr 24	Mk IV Ride Tests - Derby - Crewe - Carlisle - Crewe – Derby, TOE + 12478 + 12400 + 12300 + 10305 + 82220 (11)	47971 & 90035	[SV]
Apr 26	Mk IV Ride Tests - Derby - Crewe - Carlisle - Crewe - Derby, TOE + 12478 + 12400 + 12300 + 10305 + 82220 (12)	47972 & 90017	[SV]
Apr 27	Mk IV Ride Tests - Derby - Doncaster - Kings Cross - Derby, TOE + 12478 + 12400 + 12300 + 10305 + 82220 (13)	47972 & 91012	[SV]

Diary1990	Test Description	Loco	Test Car
Apr 30	Mk IV Ride Tests - Derby - Crewe - Carlisle - Crewe - Derby, TOE + 12478 + 12400 + 12300 + 10305 + 82220 (14)	47972 & 90021	[SV]
May 1	Mk IV Ride Tests - Derby - Doncaster - Kings Cross - Derby, TOE + 12478 + 12400 + 12300 + 10305 + 82220 (15)	47972 & Class 91	[SV]
May 9	Mk IV Ride Tests - Derby - Bedford - Derby, TOE + 12478 + 12400 + 12300 + 10305 + 82220 (16)	47973	[SV]
May 10	Mk IV Ride Tests - Derby - Doncaster - Peterborough - Derby, TOE + 12478 + 12400 + 12300 + 10305 + 82220 (17)	47971	[SV]
May 11	Mk IV Ride Tests - Derby - Doncaster - Kings Cross - Derby, TOE + 12478 + 12400 + 12300 + 10305 + 82220 (18)	47973 & 91005	[SV]
May 14	Mk IV Ride Tests - Derby - Doncaster - Kings Cross - Derby, TOE + 12478 + 12400 + 12300 + 10305 + 82220 (19)	47973 & 91005	[SV]
May 15	Mk IV Ride Tests - Derby - Doncaster - Kings Cross - Derby, TOE + 12478 + 12400 + 12300 + 10305 + 82220 (20)	47973 & 91012	[SV]
May 16	Mk IV Ride Tests - Derby - Doncaster - Kings Cross - Derby, TOE + 12478 + 12400 + 12300 + 10305 + 82220 (21)	47973 & 91002	[SV]
May 17	Mk IV Ride Tests - Derby - Doncaster - Peterborough - Derby, TOE + 12478 + 12400 + 12300 + 10305 + 82220 (22)	47972	[SV]
May 18	Mk IV Ride Tests - Derby - Doncaster - Kings Cross - Derby, TOE + 12478 + 12400 + 12300 + 10305 + 82220 (23)	47973 & Class 91	[SV]
May 20	Mk IV Transit - Derby - Bounds Green, TOE + 12478 + 12400 + 12300 + 10305 + 82220 (24)	47531	[SV]
May 21	Mk IV Ride Tests - Kings X - York - Kings X, TOE + 12478 + 12400 + 12300 + 10305 + 82220 (25)	91013	[SV]
May 22	Mk IV Ride Tests - Kings X - York - Doncaster - Derby, TOE + 12478 + 12400 + 12300 + 10305 + 82220 (26)	91013 & Class 47	[SV]
May 23	Mk IV Ride Tests - Derby - Doncaster - Kings Cross - Derby, TOE + 12478 + 12400 + 12300 + 10305 + 82220 (27)	47972 & 91013	[SV]
May 24	Mk IV Ride Tests - Derby - Doncaster - Kings Cross - Derby, TOE + 12478 + 12400 + 12300 + 10305 + 82220 (28)	47972 & 91013	[SV]
June 11	BGA - 13m Steel Carrier 961002 - Static Delta Q/Q Tests - RTC	-	-

Diary1990	Test Description	Loco	Test Car
June 12	BHA - 16m Steel Carrier 962000 - Static Delta Q/Q Tests - RTC	-	
June 16	Speno RPS-32 Rail Grinder (DR79226) repeat bogie rotation tests - RTC	-	-
June 18	BJA Steel Carrier 963000 - static torsional stiffness ($\Delta Q/Q$) tests – RTC	-	-
June 21	O&K Box (PHA) Hopper (JHA) Wagons (DB Type 25-100 2m bogies) - Set-up of ride test equipment in a 20 tonne Brake Van at Westbury Yard CE Depot	-	-
June 22	O&K Box (PHA)/Hopper (JHA) Wagons (DB Type 25-100 2m bogies) - Test run Westbury - Salisbury - Westbury	56055	Brake Van
June 23	O&K Box (PHA)/Hopper (JHA) Wagons (DB Type 25-100 2m bogies) - Test run Westbury - Salisbury - Westbury	56055	Brake Van
June 24	O&K Box (PHA)/Hopper (JHA) Wagons (DB Type 25-100 2m bogies) - Test run Westbury - Salisbury - Westbury	56055	Brake Van
July 2	Plasser 4 tonne Crane - LDRP96513 - Ride Test - Derby - Eggington - Uttoxeter - Derby, the Class 31 subsequently failed at Egginton Junction, assistance from 20170+20084 to return to Derby	31420 Assist-20170 + 20084	[TC1]
July 4	Plasser 4 tonne Crane - LDRP96513 - Ride Test - Derby - Crewe - Derby, test running up to 60 mile/h	-	[TC1]
July 9	BREL Coach 99529 (T4 Bogie) Tests Derby (Lit Lane) - Crewe - Tebay - Crewe, test running up to 110 mile/h on the WCML	47973 & 90041	[TC10]
July 10	BREL Coach 99529 (T4 Bogie) Tests Crewe (Gresty Lane) - Carlisle - Crewe - TC10 fuelled at Crewe Diesel depot, test running up to 110 mile/h on the WCML	31530 & 87013	[TC10]
July 11	BREL Coach 99529 (T4 Bogie) Tests Crewe (Gresty Lane) - Carlisle (Upperby Yard) - Crewe, test running up to 110 mile/h on the WCML	87013	[TC10]
July 12	BREL Coach 99529 (T4 Bogie) Tests Crewe (Gresty Lane) - Carlisle (Upperby Yard) - Crewe, test running up to 110 mile/h on the WCML	90020	[TC10]
July 13	BREL Coach 99529 (T4 Bogie) Tests Crewe (Gresty Lane) - Carlisle - Crewe - Derby, test running up to 110 mile/h on the WCML	90015 & 47973	[TC10]
July 16	RFS Steel Carrier wagon Ride / Stress Tests - Derby - Bedford – Derby	47974	[TC1]

Diary1990	Test Description	Loco	Test Car
July 17	ARC 102 tonne Hopper Wagon fitted with LTF25 Bogies, static torsional stiffness ($\Delta Q/Q$) and Bogie Rotation Tests – RTC		
July 18	RFS Steel Carrier wagon Ride / Stress Tests - Derby - Cricklewood – Derby	47974	[TC1]
July 23	ARC 102 tonne Hopper Wagon fitted with LTF25 Bogies Ride Test, Derby - Bedford - Loughborough (where the Wagon was loaded) - Derby. The test run was limited to 60 mile/h.	47971	[TC1]
July 24	ARC 102 tonne Hopper Wagon Bogie rotation Test - RTC	-	-
July 25	ARC 102 tonne Hopper Wagon Ride Test, Derby - Bedford – Derby	47976	[TC1]
July 29	Powell Duffryn FLA Low Container Flat Wagons LTF13 [FBT64] 1.8m Bogies Static Tests - 608001/2 - RTC	-	-
Aug 6	Powell Duffryn FLA Low Container Flat Wagons LTF13 [FBT64] 1.8m Bogies (Tare) Ride Test up to 75 mile/h, Derby - Luton - Derby- the test was delayed slightly at Flitwick due to a problem with the ZXA Van doors coming open after exit of Flitwick Tunnel	47971	[TC1]
Aug 9	Powell Duffryn FLA Low Container Flat Wagons LTF13 [FBT64] 1.8m Bogies (Part Laden) Ride Test, Derby - Luton - Derby - Train Failure (on other line) at Kettering 1 hr delay	47974	[TC1]
Aug 13	Powell Duffryn FLA Low Container Flat Wagons LTF13 [FBT64] 1.8m Bogies (Laden) Ride Test, Derby - Bedford - Derby	47974	[TC1]
Aug 14	Static Brake Pad force tests on PD Lowliner LTF13 [FBT64] 1.8m Bogies FLA wagons - RTC	-	-
Aug 15	Powell Duffryn FLA Low Container Flat Wagons LTF13 [FBT64] 1.8m Bogies (Laden) Slip/brake Test - Crewe – Winsford	47974	[TC2]
Aug 16	PD FLA Lowliner (LTF13 [FBT64] 1.8m Bogies) handbrake tests – RTC	-	-
Aug 17	Powell Duffryn FLA Low Container Flat Wagons LTF13 [FBT64] 1.8m Bogies (Tare) Slip/brake Test - Crewe – Winsford	47972	[TC2]
Aug 21	Autic Car Carrier (4289 001 6) Static Brake Tests - RTC	-	-
Aug 22	Autic Car Carrier (4289 001 6) Slip/brake Test Tare - Crewe - Winsford	-	[TC2]
Sept 17	Multimodal Wagon Ride Test (Empty Trailer) - Derby - Cricklewood - Derby	47971	[TC1]
Sept 20	Multimodal Wagon Ride Test (Tare) - Derby - Cricklewood – Derby	47520	[TC1]

Diary1990	Test Description	Loco	Test Car
Sept 25	Multimodal Wagon Ride Test (Laden) - Derby - Cricklewood – Derby	47528	[TC1]
Oct 1	Multimodal Wagon Slip/brake Test (Laden) - Crewe - Winsford, Test Car 2 generator problems during test run	47973	[TC2]
Oct 9	SNCF Car Carrier - Ride Test (Tare) Derby - Bedford – Derby	47972	[TC1]
Oct 11	SNCF Car Carrier - Ride Test (Laden) Derby - Bedford - Derby, Load shifted during test run	47973	[TC1]
Oct 12	SNCF Car Carrier - Ride Test (Laden) Derby - Bedford - Derby - Brakes binding during test run	47972	[TC1]
Oct 15	SNCF Car Carrier -Slip/brake (Laden) tests up to 60 mile/h between Crewe and Winsford on the down slow line, transit hauled by 2 x Class 20 locomotives, the test used a Class 47	20154 + 20131 47971	[TC2]
Oct 16	Static Handbrake Test on SNCF Car Carrier	-	-
Oct 18	SNCF Car Carrier -Slip/brake (Tare) tests up to 60 mile/h between Crewe and Winsford on the down slow line, transit hauled by Class 31, test locomotive Class 47	31970 47974	[TC2]
Oct 19	Static Delta Q/Q Tests - SNCF Car Carrier	-	-
Oct 22	Class 47 + 6 x Mk1 Coaches, braking tests Derby - Leamington - Didcot - Reading - Derby (late departure due to locomotive low water problem)	47844	-
Oct 24	Class 47 + 6 x Mk1 Coaches, braking tests Derby - Carstairs – Derby	47844	-
Oct 25	PD Low Platform Wagon (LTF13 [FBT64] 1.8m Bogies) static bogie rotation tests RTC	-	-
Nov 2	Multimodal Wagon handbrake test – RTC	-	-
Nov 5	Multimodal Wagon static Delta Q/Q tests	-	-
Nov 6	PD Low Platform Wagon (LTF13 [FBT64] 1.8m Bogies)static torsional stiffness (ΔQ/Q) tests	-	-
Nov 19	PD Low Platform Wagon Ride Tests (LTF13 [FBT64] 1.8m Bogies) (Tare) - Derby - Cricklewood - Derby, Loco	47973	[TC1]
Nov 20	PD Wagon(LTF13 [FBT64] 1.8m Bogies) static bogie rotation tests RTC	-	-
Nov 21	PD Low Platform Wagon (LTF13 [FBT64] 1.8m Bogies) Ride Tests (Laden) - Derby - Cricklewood - Derby	Class 47	[TC1]
Nov 26	PD Low Platform Wagon (LTF13 [FBT64] 1.8m Bogies) Ride Tests (Part Laden) - Derby - Cricklewood - Derby	47974	[TC1]
Dec 6	HAA Wagon Slip/brake Tests - Crewe - Winsford	47971	[TC2]
Dec 14	HAA Wagon Static Brake Tests, including handbrake – RTC	-	-

10. TESTING DIARY - 1991

Diary1991	Test Description	Loco	Test Car
Jan 22	Plasser SSP500 Ballast Regulator DR77701, Test Run - Derby-Old Dalby-Manton-Derby [Self Power] up to 60 mile/h, depart 09:55,arrive back at RTC 16:55	-	-
Jan 23	Plasser SSP500 Ballast Regulator DR77701, Handbrake Test	-	-
Jan 25	Marcon Hopper Wagon MAR17731, first of a new batch of PHA hopper wagons - Static torsional stiffness (ΔQ/Q) tests at RTC	-	-
Jan 28	Marcon Hopper Wagon MAR17731 - Static Bogie Rotation tests at RTC	-	-
Feb 1	Marcon Hopper Wagon rode Test (Tare) Derby-Bedford-Derby, depart RTC 09:30, arrive back 14:10	Class 47	[TC1]
Feb 7	Mk II Coaches braking tests Derby-Crewe-Carnforth-Crewe-Derby, TC2 + 5 x Mk II coaches, with one coach brakes isolated, depart RTC 07:47, arrive Crewe ETD for locomotive change 09:40, depart Crewe 10:40, Carnforth 12:05 to 13:18, arrive Crewe 14:45 locomotive change and depart 15:55, arrive back at Derby RTC 17:10	20058 + 20087 & 87023	[TC2]
Feb 11	PD Low Platform Wagon ride test, Derby - Cricklewood - Derby, 09:40 depart RTC, 15:05 arrival back	Class 47	[TC1]
Feb 15	PD Low Platform Wagon ride test, Derby - Cricklewood - Derby, 09:20 depart, 14:55 arrive back at RTC	Class 47	[TC1]
Feb 21	PGA 2-axle wagon (PR14729) ride test (Tare) Derby - Kettering - Derby, return via Loughborough for loading, 09:15 departure, 15:45 arrive back at RTC	47975	[TC1]
Feb 22	PGA 2-axle wagon (PR14729) ride test (Tare) Derby-Kettering-Derby, 09:00 departure, 13:15 arrive back at RTC	47973	[TC1]
Mar 4	Mk IV Test Train Derby-Peterborough-Bounds Green, [12224 + 12467 + 12469 + 12468 + 12323 + 12326 + 82220], testing up to 125 mile/h	47973 & 91027	[TO]
Mar 5	Mk IV Test Train, BG - Kings Cross - Peterborough - Kings Cross – BG, [12224 + 12467 + 12469 + 12468 + 12323 + 12326 + 82220], testing up to 125 mile/h	91027	[TO]
Mar 6	Mk IV Test Train, BG - Kings Cross - York - Kings Cross – BG, [12224 + 12467 + 12469 + 12468 + 12323 + 12326 + 82220], testing up to 125 mile/h	91027	[TO]
Mar 7	Mk IV Test Train, BG - Kings Cross - York - Kings Cross – BG, [12224 + 12467 + 12469 + 12468 + 12323 + 12326 + 82220], testing up to 140 mile/h	91027	[TO]

Diary1991	Test Description	Loco	Test Car
Mar 8	Mk IV Test Train, BG - Kings Cross - York - Welwyn GC – Derby, [12224 + 12467 + 12469 + 12468 + 12323 + 12326 + 82220], testing up to 140 mile/h	91027	[TO]
Mar 18	Mk IV Test Train Derby - Peterborough - Bounds Green, [12224 + 12467 + 12469 + 12468 + 12323 + 12326 + 82220], testing up to 125 mile/h	47973 & 91027	[TO]
Mar 19	Mk IV Test Train, BG - Kings Cross - York - Kings Cross - Bounds Green, [12224 + 12467 + 12469 + 12468 + 12323 + 12326 + 82220], testing up to 140 mile/h	91017	[TO]
Mar 20	Mk IV Test Train, BG - Kings Cross - York - Heaton, [12224 + 12467 + 12469 + 12468 + 12323 + 12326 + 82220], testing up to 140 mile/h	91027	[TO]
Mar 21	Mk IV Test Train, Heaton - York - Edinburgh - Heaton, 43080+43123 back 2 back, [12224 + 12467 + 12469 + 12468 + 12323 + 12326 + 82220], testing up to 125 mile/h	43080 + 43123	[TO]
Mar 22	Mk IV Test Train, Heaton - York - Edinburgh - Derby, 43080+43123 back 2 back, [12224 + 12467 + 12469 + 12468 + 12323 + 12326 + 82220], testing up to 125 mile/h	43080 + 43123	[TO]
Apr 9	ARC 102 tonne Hopper Wagon, ride test (Tare), Derby-Cricklewood-Derby (loading at Loughborough on the way back to Derby, departed RTC at 09:20, arriving back at 16:30.	47976	[TC1]
Apr 11	ARC 102 tonne Hopper Wagon, ride test (Laden), Derby-Bedford-Derby, departed RTC at 09:20, arriving back at 14:55	47976	[TC1]
Apr 16	Plasser SSP500 Test Run - Derby-Manton-Derby DR77701 [Self Power] testing up to 60 mile/h - 10:25 departure, return 15:10	-	-
Apr 28	STARFER Wagons static torsional stiffness ($\Delta Q/Q$)tests at RTC	-	-
Apr 29	RFD Container Wagons ride Test, Derby-Cricklewood-Derby, departure at 10:00, arriving back at the RTC at 15:30	47971	[TC1]
Apr 30	RFD Wagons static bogie rotation tests – RTC	-	-
May 7	RFD Container Wagons ride Test, Derby-Cricklewood-Derby, departure at 09:55, arriving back at the RTC at 15:15	47974	[TC1]
May 9	RFD Container Wagons ride Test, Derby-Cricklewood-Derby, departure at 09:15, arriving back at the RTC at 14:30	Class 47	[TC1]
May 13	RFD Container Flat Wagons, Slip/brake Test, Derby-Leicester-Nuneaton-Crewe-Winsford, delayed departure at 08:50 due to late arrival of the locomotive, arrival back at the RTC 19:15	47971	[TC2]
May 16	PD Low Platform Wagon(LTF13 [FBT64] 1.8m Bogies) ride test, Derby-Cricklewood -Derby, locomotive 47972 failed at Cricklewood, no notes about how we got rescued	47972	[TC1]

Diary1991	Test Description	Loco	Test Car
May 20	PD Low Platform Wagon(LTF13 [FBT64] 1.8m Bogies) ride test, Derby-Cricklewood -Derby	Class 47	[TC1]
May 23	Prep and travel to Frome in readiness for FY12000 tonne train trial	-	-
May 24	Merehead Quarry fitting test instrumentation to locomotives 59001, 59003 and 59005	59001 59003 59005	-
May 25	Foster Yeoman 12000tonne train trial 59005, 59001, 59003	59001 59003 59005	-
June 3	RFD Container Wagons ride re-test, Derby-Cricklewood-Derby, departed 09:30 from the RTC and return 15:30	47972	[TC1]
June 11	Plasser SP 08-275 static Delta Q/Q test – RTC	-	-
June 13	Plasser SP 08-275 Ride & Brake Tests - Derby-Manton-Derby [self-power]	-	-
June 14	Cravens Charterail Road Rail Train static torsional stiffness (ΔQ/Q) tests	-	-
June 15	Cravens Charterail Road Rail Train static torsional stiffness (ΔQ/Q) tests	-	-
June 24	Cravens Charterail Road Rail Train Tare Ride Test - Derby-Trent-Kettering-Beeston-Derby, 11:00 departure, 18:00 return back at Derby RTC	Class 47	[TC1]
July 1	Cravens Charterail Road Rail Train Laden Ride Test - Derby-Melton Mowbray, departure 08:55 however the train was terminated at Melton due to catastrophic trailer fatigue failure at Frisby. The test vehicle was detached and remained in the up loop at Melton Mowbray whilst the loco,TC1 and the ZXA returned to Derby arriving back at the RTC at 13:45	47975	[TC1]
July 14	Trailer Train II (Gloucester GPS20 1.8m bogies), static Delta Q/Q Tests – RTC	-	-
July 18	Trailer Train II (Gloucester GPS20 1.8m bogies), Ride test, Derby-Cricklewood-Derby, depart 08:55, return 15:20 at RTC	47972	[TC1]
July 29	Plasser 08-16 SPI Tamper (DR75001) static Delta Q/Q test – RTC	-	-
July 30	Multifret Wagon ride test Derby-Cricklewood-Derby, tests up to 95 mile/h, depart 10:00, arrival back 15:00 at RTC	47971	[TC1]
July 31	Trailer Train II, static torsional stiffness (ΔQ/Q) tests – RTC	-	-
Aug 15	Plasser SSP500 static torsional stiffness (ΔQ/Q) test – RTC	-	-
Sept 3	SNCF Car Carrier static torsional stiffness (ΔQ/Q) tests –RTC	-	-

Diary1991	Test Description	Loco	Test Car
Sept 10	SNCF Car Carrier ride test - Derby-Cricklewood-Derby, depart 08:45, return 15:10 at RTC	Class 47	[TC1]
Sept 12	SNCF Car Carrier ride test - Derby-Bedford-Derby, depart 09:30, return 15:30 at RTC	Class 47	[TC1]
Sept 17	Tiphook Pocket Wagon (Tare) ride test - Derby-Cricklewood-Derby, 09:00 departure, 15:30 arrival back at RTC	Class 47	[TC1]
Sept 20	Tiphook Pocket Wagon (with Trailer) ride test - Derby-Cricklewood-Derby, 09:15 departure, 15:00 arrival back at RTC	47973	[TC1]
Sept 23	Tiphook Pocket Wagon (Laden) ride test - Derby-Cricklewood-Derby, 08:50 departure, 15:15 arrival back at RTC	47831	[TC1]
Sept 25	Tiphook Pocket Wagon (Laden) Slip/brake Test Crewe-Winsford, depart 09:00 from RTC and arrival back at 18:45	47974	[TC2]
Sept 26	Kings Cross to Edinburgh Waverley, Record breaking non-stop run Class 91 + 5 x Mk IV Coaches + DVT (1T01) dep KX 09:00, 3hs 29mins record.	91012	-
Oct 8	SNCF Car Carrier static torsional stiffness ($\Delta Q/Q$) tests – RTC	-	-
Oct 15	SNCF Car Carrier ride test, Laden, Derby-Cricklewood-Derby, 09:00 depart, 15:30 arrival back at RTC	47973	[TC1]
Oct 17	SNCF Car Carrier ride test, Tare, Derby-Cricklewood-Derby, 09:00 depart, 15:30 arrival back at RTC.	47973	[TC1]
Oct 18	Multimodal Wagon static torsional stiffness ($\Delta Q/Q$) tests – RTC		
Oct 23	STARFER Wagon ride test Tare, Derby-Bedford-Derby, 09:00 departure, 15:00 arrival back at RTC	47975	[TC1]
Oct 25	STARFER Wagon ride test Laden , Derby-Bedford-Derby, 08:45 departure, 14:30 arrival back at RTC	Class 47	[TC1]
Nov 6	UIC 2-axle Tiger Tank Wagon (23 70 739 0 635-4) Ride Test, tare condition, Derby-Bedford-Derby, depart RTC at 09:30, and arrival back at 15:30	47843	[TC1]
Nov 8	UIC 2-axle Tiger Tank Wagon (23 70 739 0 635-4) Ride Test, laden condition, Derby-Bedford-Derby, depart 08:50, arrival back at RTC 14:30	47458	[TC1]
Nov 11	MOD Warflat wagons (MODA95286 + MODA95287), Impact tests at Bicester Military depot for assessment of securing method for military vehicles	-	-
Nov 12 - 13	Mk IV Coach static Sway tests RTC – 12525	-	-
Nov 18	RFS Steel Carriers, ride and axle bending stress testing, Derby-Cricklewood-Derby	47973	[TC1]

Diary1991	Test Description	Loco	Test Car
Nov 19	Mentor Test support Derby-Crewe-Warrington-Stafford-Warrington-Crewe	47974	[MEN]
Nov 25	2-axle Cement Wagons (BCC10694 + BCC11067) ride test, Derby-Bedford-Derby, depart RTC 09:10, arrival back at 14:10	47973	[TC1]
Dec 11 - 12	Mk IV Coach static Sway tests RTC – 12525	-	-
Dec 13	Mk IV Coach static Sway tests RTC - 12525, airbag came adrift during the test at 12" lift, test aborted	-	-
Dec 14	Mk IV Coach static Sway tests RTC – 12525	-	-
Dec 16	PD Low Platform Wagon ride tests, Derby-Kettering-Derby, via Manton Junction and Corby in both directions, delayed departure from RTC 10:40, arrival back 14:30	47971	[TC1]
Dec 19	Mk IV Coach static Sway tests RTC – 12525	-	-

11. TESTING DIARY - 1992

Diary 1992	Description	Locos	Test Car
Feb 6	Tiphook Multimodal Wagon - Ride Test - Derby-Cricklewood-Derby, departed RTC at 09:00, arrival back at 16:30	47973	[TC1]
Feb 13	Blue Circle Cement Wagons BCC10694 + BCC11067 (reduced Friction Damping on wagon 10694), Ride Test - Derby-Bedford-Derby, 10:25 depart, and returning back at 16:35	47971	[TC1]
Feb 27	RFS Steel Carrier Wagons - Ride Test - Derby-Bedford-Derby, departing 10:10, arrival back at RTC at 15:10	47974	[TC1]
Feb 28	Class 165 Unit static torsional stiffness ($\Delta Q/Q$) test –RTC	-	-
Mar 2	Class 165 static torsional stiffness ($\Delta Q/Q$) test –RTC	-	-
Mar 3	Class 158 (Scrubber Brake Trial) - preparation for test including generator installation	158775	-
Mar 4 - 5	Class 158 (Scrubber Brake Trial) - preparation for test strain gauge bolts installation	158775	-
Mar 9	Blue Circle Cement Wagon (reduced Friction Damping) - static torsional stiffness ($\Delta Q/Q$) test –RTC	-	-
Mar 10 - 13	Class 158 (Scrubber Brake Trial) - preparation for test	158775	-
Mar 19 - 20	Class 158 (Scrubber Brake Trial) - static loading tests	158775	-
Mar 23	Class 158 (Scrubber Brake Trial) - Derby - Old Dalby – Derby, departure 09:30, arriving back at the RTC at 15:30	158775	-
Mar 24	Class 158 (Scrubber Brake Trial) - Derby - Liverpool – Derby, departing at 08:30 and arriving back at the RTC at 15:00	158775	-
Mar 25	Class 158 (Scrubber Brake Trial) - Derby - Norwich – Derby, departing at 08:30 and arriving back at the RTC at 16:15	158775	-
Mar 26 - 27	Class 158 (Scrubber Brake Trial) - Derby - Old Dalby – Derby, departing at 08:30 and arriving back at the RTC at 16:30	158775	-
Mar 28	Class 158 (Scrubber Brake Trial) - de-rigging / instrumentation removal at the RTC	158775	-
Apr 13	HAA Wagons Coal Train - coupling stresses and coupling shocks measurements - Toton - Melton - Asfordby (2 x Class 58 locomotives)	Class 58 + Class 58	-
May 7	Permaquip Wagon - Wedge Tests – RTC	-	-
May 20	Hot Axlebox investigation tests (TEA Tank Wagon),Derby - Bedford - Derby, departed at 09:00 and return 15:30	20132 + 20090	[TC1]
June 2	SNCF 2-axle Car Carrier Wagon Ride Tests, Derby - Bedford - Derby, depart 09:15, returned at 13:50	Class 47	[TC1]
June 3	SNCF 2-axle Car Carrier Wagon Ride Tests, Derby - Cricklewood - Derby, depart 09:11, returned at 15:20	Class 47	[TC1]

Diary 1992	Description	Locos	Test Car
June 4	SNCF 2-axle Car Carrier Wagon Ride Tests, Derby - Bedford - Derby, depart 09:15, returned at 14:15	Class 47	[TC1]
June 12	Preparations at Mountsorrell for stress tests on Redlands Self Discharge Train wagons	-	-
June 13	Preparations at Mountsorrell for stress tests on Redlands Self Discharge Train wagons, instrumentation fitted in the rear cab of 56049	56049	-
June 14	Preparations at Mountsorrell for stress tests on Redlands Self Discharge Train wagons, instrumentation fitted in the rear cab of 56049	56049	-
June 15	Redlands Self Discharge Train test run (measurements of stress in the wagon body structure) Mountsorrell - Cricklewood, 56049 + 56062, departed 03:50 arrived Cricklewood at 07:30	56049 + 56062	-
June 16	SNCF Container Flat Wagon, Ride Test (Tare + Ballast Load + Tight Coupled) - Derby - Bedford – Derby, 09:00 depart, 14:30 return arrival at RTC	47976	[TC1]
June 17	SNCF Container Flat Wagon, Ride Test (Tare + Ballast Load - Loose Coupled) - Derby - Bedford - Derby, 08:55 depart, 15:15 return arrival at RTC	47975	[TC1]
June 19	SNCF Container Flat Wagon Ride Test (Tare condition - Tight Coupled) - Derby - Bedford - Derby, 08:55 depart, 13:50 return arrival at RTC	47975	[TC1]
June 22	SNCF Container Flat Wagon, Ride Test (Loaded condition - Tight Coupled) - Derby - Bedford - Derby, 08:55 depart, 13:40 return arrival at RTC	47973	[TC1]
June 30	SNCF Container Flat Multimodal Wagon, Ride Test - Derby-Cricklewood-Derby, depart at 09:00, arriving back at the RTC at 15:20	47972	[TC1]
July 3	Preparations for testing on Bardon Hill Quarries supertrain (Class 60)	60011	-
July 4	Bardon Hill Quarries Class 60 Super Train Test, Bardon Hill to Acton via West Hampstead, [Data Loggers], dep Bardon Quarry 09:15, arrive Acton 15:05	60011	-
July 23	Prorail Nuclear Flask wagon Slip/brake Tests, Crewe-Winsford-Warrington	47973	[TC2]
Aug 3 - 5	Cravens Charterail Road Rail Train Strain Gauge installation	-	-
Aug 6	Cravens Charterail Road Rail Train (Laden) Ride Test - Derby-Bedford-Derby, depart 09:20, arrive back at RTC at 14:45	47972	[TC1]
Aug 7	Cravens Charterail Road Rail Train, static brake tests - RTC	-	-
Aug 10	Cravens Charterail Road Rail Train (Laden) Slip/brake test, Crewe-Winsford, depart 08:15, arrive back at RTC at 18:30	47976	[TC2]

Diary 1992	Description	Locos	Test Car
Sept 9	Powell Duffryn Low Platform Wagons - Ride Test, Derby-Cricklewood-Derby, departed Derby 08:35, return 16:10	47973	[TC1]
Sept 14	NSE Skako Hopper Wagons - Ride Test, Derby-Bedford-Derby, 09:15 departed, and arrived back at RTC at 14:30	47971	[TC1]
Oct 6 - 9	Class 90 Locomotives - bogie frame fractures investigation, set-up instrumentation / strain gauge fitting - Crewe EMD	90134	-
Oct 13	Class 90 Locomotives - bogie frame fractures investigation, test run - Crewe - Carlisle – Crewe	90134 + 90019	-
Oct 14	Class 90 Locomotives - bogie frame fractures investigation, test run - Crewe - Stafford - Crewe	90134 + 90144	-
Oct 26	Cravens Charterail Road Rail Train - torsional stiffness (ΔQ/Q) test - RTC	-	-
Oct 27	Cravens Charterail Road Rail Train - Bogie Rotation static tests - RTC	-	-
Oct 28 - 29	Cravens Charterail Road Rail Train - instrumentation preparation - RTC	-	-
Oct 30	Cravens Charterail Road Rail Train - Brake Block load measurements - RTC	-	-
Nov 3	Cravens Charterail Road Rail Train - Ride Test (Tare) - Derby - Bedford - Derby, depart at 09:00, returning at 14:50	47971	[TC1]
Nov 5	Cravens Charterail Road Rail Train - Slip/brake test (Tare) - Crewe-Winsford, depart at 08:20, returning at 18:30	47976	[TC1]
Nov 11	MOD Bicester Military sidings - Impact Tests (securing of military vehicles onto PFA 'Warflat' Bogie Flat Wagons - UIC Tests comparison of chains and span-set strap methods) - Test instrumentation installed in Mitsubishi Shogun (E306LRF)	MOD Shunter	-
Nov 17	2-axle Cement Wagons, BCC10974, BCC10682, APCM9196, APCM9145, BCC10950, modified friction damping, static torsional stiffness (ΔQ/Q) tests - RTC	-	-
Nov 18	Preparations for Mk IV testing at Craigentinny depot (Edinburgh)	-	-
Nov 19	Mk IV Test run P8 vs. P10 wheel profiles assessment - Craigentinny - Edinburgh - Doncaster - Edinburgh – Craigentinny, 91031 + 12211 + 12525(P8) + 12428 + 10312 + 11223 + 11222 (P10) + 82224. (test cancelled due to 90031 locomotive failure)	90031	[TO]
Nov 20	Mk IV Test run P8 vs. P10 wheel profiles assessment, Craigentinny - Edinburgh - Doncaster - Edinburgh – Craigentinny, 91007 + 12211 + 12525(P8) + 12428 + 10312 + 11223 + 11222 (P10) + 82224	91007	[TO]

Diary 1992	Description	Locos	Test Car
Nov 23	Tiphook KPA Wagon (Tare) - Slip/brake Test - Crewe-Winsford, Warrington, 08:30 departure, and 18:05 return to RTC	47975	[TC2]
Nov 26	Tiphook KPA Wagon (Laden) - Slip/brake Test - Crewe-Winsford, Warrington, 08:10 departure, and 18:00 return to RTC	47972	[TC2]
Dec 17	Tiphook KPA Wagon (Tare) - Slip/brake Test- Crewe-Winsford, Warrington, 08:30 departure, and 18:30 return to RTC	47971	[TC2]

12.   TESTING DIARY - 1993

Diary 1993	Description	Loco	Test Car
Jan 11	Class 507 Unit static torsional stiffness (ΔQ/Q) test - RTC	-	-
Jan 14	MOD Bicester Military sidings - Impact Tests (securing of military vehicles onto freight wagons) - Test instrumentation installed in the Test Section Sherpa Van	-	-
Jan 18	Iron Ore Tippler Wagons (Laden) Slip/brake Test - Crewe-Winsford-Warrington	-	[TC2]
Jan 21	Iron Ore Tippler Wagons (Laden) Slip/brake Test - Crewe-Winsford-Warrington, 08:15 dep Derby, return 17:45	47971	[TC2]
Jan 26	Iron Ore Tippler Wagons - Ride Test incl. Track Assessment - Derby-Kettering-Derby, 08:50 dep Derby, Arr Kettering 11:15, return at Derby 13:30	-	[TC1]
Feb 2	Iron Ore Tippler Wagons (Tare) Slip/brake Test - Crewe-Winsford-Warrington, 08:40 dep Derby, return 18:00, 12 Slip Tests Completed (Dry)	47972	[TC2]
Feb 4	Iron Ore Tippler Wagons (Tare) Slip/brake Test (including Ride Assessment) - Crewe-Winsford-Warrington, transit locomotive 47973, test locomotive 47971, 08:40 dep Derby, return 18:45, problems with Vac Brake on 47971 during testing	47973 & 47971	[TC2]
Mar 5	Kershaw Ballast Regulator, (DR77501) self-propelled Ride / Brake test run, Derby-Manton-Derby, dep 08:30, return 11:22 (6Z09)	-	-
Mar 15	Blue Circle Cement Wagons (BCC10694 + BCC10682) static torsional stiffness (ΔQ/Q) tests - RTC	-	-
Mar 18	Blue Circle Cement Wagons (BCC10694 + BCC10682) Laden Ride Test, Derby - Bedford - Derby, 10:20 depart, 15:10 return at Derby	47824	[TC1]
Mar 19	Blue Circle Cement Wagons (BCC10694 + BCC10682) Static Delta Q/Q Tests – RTC	-	-
Mar 24	Tiphook 82 tonne Container Flat Wagon - static torsional stiffness (ΔQ/Q) and Bogie Rotation tests - RTC	-	-
Mar 29	Tiphook 82 tonne Container Flat Wagon - static tests (Brake Block Forces) – RTC	-	-
Mar 30	Tiphook 82 tonne Container Flat Wagon - Slip/brake Test - Crewe-Winsford-Warrington, late departure 11:00, back at Derby 19:00	47975	[TC2]
Apr 1	Cravens Charterail Road Rail Train (Laden) - Slip/brake Test - Crewe-Winsford-Warrington, 09:00, back at Derby RTC 18:30	47975	[TC2]

Diary 1993	Description	Loco	Test Car
Apr 6	Tiphook 82 tonne Container Flat Wagon (Laden) - Slip/brake Test - Crewe-Winsford-Warrington, 08:25, back at Derby 18:30, problems noted with 47972 Vacuum Brake chamber and locomotive underpowered, 75mile/h testing could not be achieved as planned	47972	[TC2]
Apr 14	RFS Intermodal Wagon static torsional stiffness ($\Delta Q/Q$) and Bogie Rotation tests– RTC	-	-
Apr 21	RFS Intermodal Wagon (Tare) Ride tests - Derby-Cricklewood-Derby depart 08:40, return 15:20	Class 47	[TC1]
Apr 26	Conaco 90 tonne Tank Wagon - Strain Gauging preparations – RTC	-	-
Apr 27	RFS Intermodal Wagon (Part Laden) Ride tests - Derby-Cricklewood-Derby depart 09:20, return 15:40	Class 47	[TC1]
Apr 30	RFS Intermodal Wagon (Laden) Ride tests - Derby-Cricklewood-Derby, depart 08:40, return 15:00	47971	[TC1]
May 9	Slip/brake testing - Crewe-Winsford (no record of details)	-	[TC2]
May 11	Conaco 90 tonne TDA (854xx) Tank Wagon - Ride Test and Strain Gauging preparations - RTC	-	-
May 12	Conaco 90 tonne TDA (854xx) Tank Wagon - Ride Test - Derby-Bedford-Derby, 09:50 depart, 15:30 return to RTC	47973	[TC1]
May 14	Conaco 90 tonne TDA (854xx) Tank Wagon - Ride Test - Derby-Bedford-Derby, 08:30 depart, 14:30 return to RTC	47973	[TC1]
May 19 - 20	Arbel-Fauvet Rail (WIA) Car Carrier Wagon 85.70.4971.000 5 wagon articulated set (Y39 2.2m bogies)- preparations and instrument fitting for Ride Test	-	-
May 21	Arbel-Fauvet Rail (WIA) Car Carrier Wagon 85.70.4971.000 5 wagon articulated set (Y39 2.2m bogies) -Ride Test, Derby-Cricklewood-Derby	47802	[TC1]
May 27	Arbel-Fauvet Rail (WIA) Car Carrier Wagon 85.70.4971.000 5 wagon articulated set (Y39 2.2m bogies) - Ride Test, Derby-Cricklewood-Derby	47973	[TC1]
May 29	Shell 2-axle Tank Wagon test preparations at Tyne Yard, instrumentation installation in rear cab of Class 37 locomotive	-	-
May 30	Shell 2-axle Tank Wagon - Test Run - Tyne Yard - York - Wakefield - Stockport - Chester - Hooton	Class 37	-
May 31	Shell 2-axle Tank Wagon instrumentation removal at Hooton	-	-
June 7	HOBC (High Output Ballast Cleaner) DR76101 - Static Bogie Rotation Tests – RTC	-	-
June 8	HOBC (High Output Ballast Cleaner) DR76101 - static torsional stiffness ($\Delta Q/Q$) tests – RTC	-	-

Diary 1993	Description	Loco	Test Car
June 15	Freightliner Wagons (Laden condition, non-Asbestos Brake Pads performance trials), Slip/brake tests, Crewe-Winsford, Warrington, 09:30 depart Derby, 18:15 return	47971	[TC2]
June 23	HOBC (High Output Ballast Cleaner) DR76101 - Ride Tests, Derby-Cricklewood-Derby, depart Derby 09:30, return 15:45	47972	[TC1]
June 24	HOBC (High Output Ballast Cleaner) DR76101 - Braking Tests, Derby-Leicester-Derby (3 round trips), depart Derby 08:15, return 15:15	47972	-
July 6	Channel Tunnel ET Locomotive 0032 testing at Old Dalby, transit out Derby-Old Dalby with TC1+TC2	47972	[TC1] + [TC2]
July 7	Channel Tunnel Locomotive testing at Old Dalby, with TC1+TC2	47972	[TC1] + [TC2]
July 11	RFS Intermodal Wagon (Laden) Ride tests - Derby-Cricklewood-Derby, depart 08:40, return 15:00	47971	[TC1]
July 13	RFS Intermodal Wagon (Laden) Ride tests - Derby-Bedford-Derby, depart 10:40, return 13:45	47971	[TC1]
July 14 - 15	Mk IV Coach (ABB T4-5 Bogies 10310 [SV]) static Sway Testing – RTC	-	-
July 16 - 23	Mk IV Coach (ABB T4-5 Bogies 10310 [SV]) - Temporary Test Car and instrumentation set-up - RTC	-	-
July 26	Mk IV Coach (ABB T4-5 Bogies 10310 [SV]) - Ride Test - Instrument proving and calibration test runs - Derby-Leicester- Derby	47972	-
July 27	Mk IV Coach (ABB T4-5 Bogies 10310 [SV]) - Ride Test - Instrument proving and calibration test runs - Derby-Leicester- Derby	47976	-
July 28	Mk IV Coach (ABB T4-5 Bogies 10310 [SV]) - Ride Test - Instrument proving and calibration test runs - Derby-Leicester- Derby	47973	-
July 29	Mk IV Coach (ABB T4-5 Bogies 10310 [SV]) - Ride Test - Derby-Stoke-Bletchley-Nuneaton-Bletchley-Rugby-Stoke-Derby, depart Derby 08:55, return 19:30	47973	-
July 30	Mk IV Coach (ABB T4-5 Bogies 10310 [SV]) - Ride Test - Derby-Stoke-Bletchley-Nuneaton-Bletchley-Rugby-Stoke-Derby, depart Derby 08:30, return 19:00	47973	-
July 31	Mk IV Coach (ABB T4-5 Bogies 10310 [SV]) Transit / Ride Test - Derby - York - Edinburgh (Craigentinny), 10:45 departure from Derby	47976	-
Aug 1	Mk IV Coach (ABB T4-5 Bogies 10310 [SV]) loading and tyre turning at Craigentinny	-	-

Diary 1993	Description	Loco	Test Car
Aug 2-3	Mk IV Coach (ABB T4-5 Bogies 10310 [SV]) testing on ECML - from Edinburgh - West Coast Main Line (UIC515 Homologation in preparation for T4-5 bogie installation on the New European Night Stock coaches	Class 47	-
Aug 4	Mk IV Coach (ABB T4-5 Bogies 10310 [SV]) Laden testing on ECML - between Edinburgh - Berwick - Darlington - York	91022	-
Aug 5	Mk IV Coach (ABB T4-5 Bogies 10310 [SV]) Tare testing on ECML - between Edinburgh - Berwick - Darlington - York	91022	-
Aug 6	Mk IV Coach (ABB T4-5 Bogies 10310 [SV]) Test Train Transit move Edinburgh (Craigentinny) to Derby	Class 47	-
Aug 17	Arbel-Fauvet Rail (WIA) Car Carrier 85.70.4971.000 5 wagon articulated set (Y39 2.2m bogies)- Ride Test, Derby-Cricklewood-Derby	Class 47	[TC1]
Aug 19	Suko 100 tonne Tank Wagon static Bogie Rotation test - RTC	-	-
Aug 23	British Rail Research UTU vehicle ride test set-up - RTC	-	-
Aug 24	British Rail Research UTU2 Unit ride test set - Derby-Melton Mowbray- Old Dalby test line (65 mile/h maximum attained during testing) UTU2 was converted in 1987 from Class 101 power-twin unit - formed - 977391 (ex-51433) + 999602 (ex-4REP EMU vehicle 62483) + 977392 (ex-53167) The UTU2 driving vehicles were unique, as they were the only first generation DMU cars to have been installed with Auto Air and EP Air Brakes		-
Aug 25	Permaquip Multicar (Road Rail Personnel Carrier Vehicle) ride / braking / stability tests at Midland Railway Centre Swanwick Junction site	-	-
Sept 1	Arbel-Fauvet Rail (WIA) Car Carrier 85.70.4971.000 5 wagon articulated set (Y39 2.2m bogies)- static tests (Bogie Rotation and Handbrake Test) Laden - RTC		
Sept 7	Arbel-Fauvet Rail (WIA) Car Carrier 85.70.4971.000 5 wagon articulated set (Y39 2.2m bogies) (Laden) Ride Test - Derby-Cricklewood-Derby, 08:40 depart Derby, return 15:00	47973	[TC1]
Sept 21	Powell Duffryn Supercube Wagon (96600) -Ride Tests (Tare) - Derby-Cricklewood-Derby, 0810 depart, 15:30 return Derby	Class 47	[TC1]
Sept 22	Powell Duffryn Supercube Wagon (96600) -wedge tests / investigations – RTC	-	-
Sept 24	Powell Duffryn Supercube Wagon (96600) -Ride Tests (Laden) - Derby-Bedford-Derby, 0825 depart, 13:30 return Derby	Class 47	[TC1]
Sept 28	Arbel-Fauvet Rail (WIA) Car Carrier Wagon 85.70.4971.000 5 wagon articulated set (Y39 2.2m bogies) - Slip/brake Tests, Crewe-Winsford-Warrington, depart 08:10, return 18:15	47975	[TC2]

Diary 1993	Description	Loco	Test Car
Oct 6	Channel Tunnel ET Locomotive 0032 testing at Old Dalby, Derby-Old Dalby with TC1+TC2, locomotive 47972		
Oct 13	USP5000C (DX77316) Ballast Regulator, Ride Test (poor rode investigation), portable equipment, from Hither Green depot.		
Nov 8 - 11	Mk IV Coach (ABB T4-5 Bogies 10310 [SV]) - test runs on ECML		
Nov 16	Class 86 (TET Tests) Crewe - Winsford	86605	
Dec 9	National Power JHA-Hopper Wagon (Powell Duffryn LTF-25 2m bogies) NP19400, Ride Test , Derby-Cricklewood-Derby, return via Mountsorrell for loading with aggregate. Depart Derby 09:15, return 18:30	47618	[TC1]
Dec 11 - 12	Class 92 (92002) Static Sway Testing - RTC Derby	92002	-
Dec 13	Class 92 (92002) Static Bogie Rotation Testing - RTC Derby	92002	-
Dec 14	National Power JHA-Hopper Wagon (Powell Duffryn LTF-25 2m bogies) NP19400Ride Test (Part Laden), Derby-Cricklewood-Derby. Depart Derby 09:00, return 15:30.	Class 47	[TC1]
Dec 16	National Power JHA-Hopper Wagon (Powell Duffryn LTF-25 2m bogies) NP19400Ride Test (Fully Laden), Derby-Bedford-Derby. Depart Derby 08:45, return 14:30	Class 47	[TC1]
Dec 17	Kershaw CCT 45/10 TRAMM DR98801 - self-powered ride test Derby - Kettering – Derby		

13. TESTING DIARY - 1994

Diary 1994	Description	Loco	Test Car
Jan 5	Blue Circle Cement Wagons (BCC10682 + BCC10694) static torsional stiffness ($\Delta Q/Q$) tests - RTC	-	-
Jan 6	National Power JHA-Wagon (Powell Duffryn LTF-25 2m bogies) - Slip/brake Tests - Crewe-Winsford, Warrington, 08:30 depart Derby, 17:45 return	47972	[TC2]
Jan 16	Long Container Train braking, static tests at Lawley St Freightliner Terminal (Birmingham)	47241	
Jan 18	National Power JHA-Wagon (Powell Duffryn LTF-25 2m bogies) - Slip/brake Tests - Crewe-Winsford, Warrington, 08:15 depart Derby, 18:00 return	47981	[TC2]
Feb 14	Class 312 Unit Axlebox temperatures investigations - Clacton - preparations for testing		
Feb 15 - 16	Class 312 Unit Axlebox temperatures investigations - Clacton - Liverpool Street - test running		
Feb 24	National Power Hopper JHA-Wagon (Laden) Ride re-test, Derby-Bedford-Derby, depart 09:10, return 14:30	Class 47	[TC1]
Feb 25	Preparations for long aggregate train braking tests - Western Region		
Feb 26	Long Aggregate Train braking tests, Merehead, instrumentation installation - Class 59 loco		
Feb 27	Long Aggregate Train braking tests, Merehead Quarry, static braking tests, Class 59	59005	
Mar 8 - 9	Class 59/2 (59201) static torsional stiffness ($\Delta Q/Q$) and bogie rotation / clearance tests / checks at RTC Derby	59201	
Mar 10	Class 59/2 (59201) instrumentation preparations	59201	
Mar 14	Class 59/2 (59201), ride test, Derby-Cricklewood-Derby, 08:55 depart, 15:45 return	59201	[TC1]
Mar 15	Class 59/2 (59201), braking test, Derby-Leicester-Derby, 10:00 depart, 15:30 return	59201	
Mar 18	Class 318 braking tests - preparations at Shields Road depot	-	-
Mar 19	Class 318 braking tests (318253) brake test running, local Glasgow routes	318253	-
Apr 7	Class 466 ASB (Advanced Suburban Bogie) static Sway Testing at RTC Derby (466028 - DTOS 79339)	-	-
Apr 8	Class 466 ASB (Advanced Suburban Bogie, 466028) static tests at RTC	-	-
Apr 11-12	Starfer Single Line Spoil Handling System Wagon (YDA-B) 92201 static tests and preparations for ride testing - RTC	-	-

Diary 1994	Description	Loco	Test Car
Apr 13	Starfer Single Line Spoil Handling System Wagon (YDA-B) 92201 (Tare) ride test - Derby-Bedford-Derby, 09:15 depart, 15:35 return	47717	[TC1]
Apr 15	Starfer Single Line Spoil Handling System Wagon (YDA-B) 92201(Laden) ride test - Derby-Bedford-Derby, 09:15 depart, 15:35 return	Class 47	[TC1]
Apr 27 - 28	British Gypsum KFA (BGL95310) Wagon static torsional stiffness (ΔQ/Q) and Bogie rotation tests– RTC		
Apr 29	British Gypsum KFA (BGL95310) Wagon preparations for Ride test – RTC		
May 3	British Gypsum KFA (BGL95310) Wagon (Tare) ride test - Derby-Cricklewood-Derby [TC1], 09:25 depart, 15:30 return	-	[TC1]
May 5	British Gypsum KFA (BGL95310) Wagon (Tare+Empty Containers) ride test - Derby-Cricklewood-Derby, 08:38 depart, 15:00 return	-	[TC1]
May 11	Class 466 ASB (Advanced Suburban Bogie) Ride Test & Bogie Stress Tests, Derby-Leicester-Derby, (466028)	Class 47	-
May 12	Class 466 ASB (Advanced Suburban Bogie) Ride Test & Bogie Stress Tests, Derby-Kettering-Derby, (466028), depart 12:45, return 14:45	47975	-
May 19	British Gypsum KFA (BGL95310) Wagon (Laden) ride test - Derby-Cricklewood-Derby, 08:45 depart, 15:25 return	Class 47	[TC1]
May 23	British Gypsum KFA (BGL95310)Wagon (Laden) Slip/brake test, Crewe-Winsford, depart Derby 08:10, return 18:30	Class 47	[TC2]
May 24	Class 59 locomotive ride investigations 59004, bogie rotation tests at RTC Derby	59004	-
May 25	Class 466 ASB (Advanced Suburban Bogie) static Sway Testing at Strawberry Hill depot (466028 - DMOS 64887)	-	-
May 26	Class 466 ASB (Advanced Suburban Bogie) static Sway Testing at Strawberry Hill depot (466028 - DMOS 64887)	-	-
May 27	Class 466 ASB (Advanced Suburban Bogie) static Sway Testing at Strawberry Hill depot (466028 - DMOS 64887)	-	-
June 3	Class 59 locomotive ride investigations 59004, ride test run Derby to Birmingham, portable ride data recorder used.	59004	-
June 8	FSA Wagons (VNH1 1.8m bogies), preparations for ride test and static tests – RTC	-	-
June 9	FSA Wagons (VNH1 1.8m bogies), ride test, Derby-Cricklewood-Derby, depart 07:45, return 15:15	47746	[TC1]
June 10	FSA Wagons (VNH1 1.8m bogies), static torsional stiffness (ΔQ/Q) and bogie rotation tests - RTC		

Diary 1994	Description	Loco	Test Car
June 11	Transfesa Wagon (TRAN93500) preparations for ride test – RTC		
June 13	Transfesa Wagon (TRAN93500) - ride test, Derby-Bedford-Derby, depart 07:25, return 14:50	47981	[TC1]
June 21	FSA Wagons (VNH1 1.8m bogies), ride testing, Derby-Sheffield-Derby-Birmingham-Derby, depart 09:45, return 15:15	47976	[TC2]
June 22	FSA Wagons (VNH1 1.8m bogies), static Bogie rotation tests – RTC		
June 23	Transfesa Wagon (Laden) - ride test, Derby-Bedford-Derby, depart 06:55, return 13:40	Class 47	[TC1]
June 30	Transfesa Wagon (Part Laden) - ride test, Derby-Bedford-Derby, depart 07:10, return 12:55	47981	[TC1]
July 14	British Gypsum KFA (BGL95310) Wagon (Tare) ride test - Derby-Cricklewood-Derby, 06:55 depart, 12:55 return	47975	[TC1]
July 18	British Gypsum KFA (BGL95310) Wagon (Part Loaded Containers) ride test - Derby-Cricklewood-Derby, 07:20 depart, 12:02 return	47981	[TC1]
July 25	British Gypsum KFA (BGL95310)Wagon (Loaded Containers) ride test - Derby-Cricklewood-Derby	47981	[TC1]
Aug 1	Transfesa Wagon(TRAN93500) ride testing - Derby - Dover - Stewarts Lane	Class 47	[TC1]
Aug 2	Transfesa Wagon (TRAN93500) ride testing - Stewarts Lane - Clapham – Derby	Class 47	[TC1]
Aug 10	Transfesa Wagon (TRAN93500) ride testing - Derby - Dover - Stewarts Lane	47973	[TC1]
Aug 11	Transfesa Wagon (TRAN93500) ride testing - Stewarts Lane - Clapham - Derby	47973	[TC1]
Aug 16	Class 37 preparations for handbrake testing at Craigentinny Depot (Edinburgh) - test required following runaway and collision damage	37113	
Aug 17	Class 37 handbrake testing at Craigentinny Depot (Edinburgh) - test required following runaway and collision damage	37113	
Sept 5	British Gypsum KFA (BGL95310) Wagon (Laden) ride test - Derby-Cricklewood-Derby, 08:05 depart Derby, 14:05 return.	47972	[TC1]
Sept 7	British Gypsum KFA (BGL95310) Wagon (Laden) ride test - Derby-Cricklewood-Derby, 07:20 depart Derby, 13:55 return.	47981	[TC1]
Sept 21	Channel Tunnel shuttle vehicles ride tests, preparations in Coquelles F40 maintenance depot (France)	9026 + 9031	-

Diary 1994	Description	Loco	Test Car
Sept 22	Channel Tunnel shuttle vehicles ride tests, preparations in Coquelles F40 maintenance depot (France)	9026 + 9031	-
Sept 23	Channel Tunnel shuttle Tourist vehicle and car carrier vehicles ride tests (Laden condition), 4 round trip runs between Coquelles & Folkestone on shuttle lines through Channel Tunnel.	9027 + 9031	-
Sept 24	Channel Tunnel shuttle Tourist vehicle and car carrier vehicles ride tests (Laden condition), 4 round trip runs between Coquelles & Folkestone on shuttle lines through Channel Tunnel.	9027 + 9031	-
Sept 25	Channel Tunnel shuttle Tourist vehicle and car carrier vehicles ride tests (Tare condition), 2 round trip runs between Coquelles & Folkestone on shuttle lines through Channel Tunnel.	9027 + 9031	-
Oct 5	Transfesa Wagon (TRAN93500) - (following suspension modifications) ride test, Derby-Bedford-Derby depart 07:30, return 13:00	Class 47	[TC1]
Oct 9 - 10	Class 320 braking tests, static brake characteristics tests & preparations at Shields depot (Glasgow)	-	-
Oct 11 - 13	Class 320 (Unit 320301) braking tests, Glasgow - Kilwinning – Ayr	-	-
Oct 15	Loading of 320301 at Shields depot	-	-
Oct 16	Class 320 braking tests, static brake characteristics tests & preparations at Shields depot (Glasgow)	-	-
Oct 17	Class 320 (Unit 320301) braking tests, Glasgow - Lanark - Shields - Yoker - Lanark – Shields	-	-
Oct 18 - 19	Class 320 (Unit 320301) braking tests, Glasgow - Kilwinning – Ayr	-	-
Oct 25	Powell Duffryn Supercube Wagon (96600) - Tare Ride Test - Derby-Cricklewood – Derby	47971	[TC1]
Oct 28	Powell Duffryn Supercube Wagon (96600) - Part Laden Ride Test - Derby-Cricklewood – Derby	Class 47	[TC1]
Nov 1	Powell Duffryn Supercube Wagon (96600) - Laden Ride Test - Derby-Cricklewood – Derby	47981	[TC1]
Nov 8	Powell Duffryn Supercube Wagon (96600) - Laden Slip/brake Test - Crewe-Winsford, depart Derby 07:55, return 18:15	Class 47	[TC2]
Nov 9	Powell Duffryn Supercube Wagon (96600) - Laden Slip/brake Test - Crewe-Winsford, depart Derby 07:55, return 18:05	Class 47	[TC2]
Nov 19 - 20	Class 92 locomotive instrumentation preparation - RTC		[TC2]

Diary 1994	Description	Loco	Test Car
Nov 21	Class 92 locomotive - transit and testing (ride, brake, air pressures, pantograph), - Derby - Crewe	Class 47 & 92003	[TC2]
Nov 22	Class 92 locomotive - WCML testing (ride, brake, air pressures, pantograph) - Crewe - Carlisle – Crewe	Class 47 & 92003	[TC2]
Nov 24	Plasser NFS Wagons static torsional stiffness ($\Delta Q/Q$) and Bogie Rotation tests– RTC	-	-
Nov 27 - 30	Class 92 locomotive instrumentation preparation - Dollands Moor	92003	[TC2]
Dec 1	Class 92 locomotive instrumentation preparation - Dollands Moor	92003 + 92018	[TC2]
Dec 12 - 15	Class 92 locomotive test running (Channel Tunnel) from - Dollands Moor, locomotives 92003 + 92018	92003 + 92018	[TC2]

14.   TESTING DIARY - 1995

Diary1995	Description	Loco	Test Car
Jan 4	Class 59 locomotive static bogie rotation tests - RTC	59004	-
Jan 17	Plasser NFS-D Wagon DR92223 (+ barrier wagon)Tare - ride test - Derby-Cricklewood - Syston	47981	[TC1]
Jan 19	Plasser NFS-D Wagon Laden - ride test - Derby-Bedford – Derby	47972	[TC1]
Jan 24	Plasser MFS-D Wagon (DR92241) static bogie rotation and delta Q/Q tests - RTC	-	-
Jan 26	Plasser MFS-D Wagon (DR92241) static bogie rotation and delta Q/Q tests - RTC	-	-
Jan 30	Plasser NFS-D Wagon DR92223 (laden condition + barrier wagon), Slip/brake test, Crewe-Winsford, Warrington, late departure due to generator problems on test car 2, depart 09:15, return 18:15	47972	[TC2]
Feb 1	Plasser NFS barrier wagon only, Slip/brake test, Crewe-Winsford, Warrington, depart 08:20, return 18:15	47972	[TC2]
Feb 7	Plasser MFS-D wagon (DR92241) ride test - Derby-Cricklewood-Derby, return via Redland Mountsorrell for loading	-	[TC2]
Feb 25	Loram Rail Grinder DR79200 - static tests, bogie rotation tests, parking brake and block force measurements - RTC	-	-
Mar 5 - 6	Class 60 + HAA Wagon braking tests from Toton to Didcot	Class 60	-
Mar 13	Class 312 EMU (312723) braking tests set-up at Ilford depot	-	-
Mar 14 - 16	Class 312 EMU (312723) braking tests - Colchester – Shenfield	-	-
Mar 26	Class 321/9 EMU (321902) preparations for testing at Leeds Neville Hill (braking tests)	-	-
Mar 27	Class 321/9 EMU (321902) testing from Leeds Neville Hill - Doncaster (braking tests)	-	-
Mar 29	Class 325 testing (braking tests) - Warrington-Winwick	-	-
Mar 30	Class 325 testing (braking tests) - Warrington-Winwick	-	-
Apr 10	Plasser NFS-D Wagon DR92223 (Tare condition + barrier wagon), Slip/brake test retest, Crewe-Winsford, Warrington	Class 47	[TC2]
May 1	National Power JMA Coal Wagon NP19601 - static tests (bogie rotation) - RTC	-	-

Diary1995	Description	Loco	Test Car
May 5 – 9	Class 60 drawgear investigations - Toton (preparations for testing)	Class 60	-
May 10 - 11	Class 60 drawgear investigations - Toton - Luton - Toton test running	Class 60	
May17	National Power JMA Coal Wagon NP19601- ride test, Derby-Bedford-Derby	47981	[TC1]
May 23 - 24	Manchester Metrolink tram vehicles - ride investigations, testing over all Metrolink routes	-	-
May 29	Peterborough 150, preparations of ride comfort test equipment at Derby	-	-
May 30	Peterborough 150, installation of the ride quality and comfort test equipment, vehicle 12204 at Bounds Green depot	-	-
May 31	Peterborough 150, transit movement from Bounds Green to Heaton depot, including braking tests, formation 82231, 11275, 11274, 10323, 11273, 12204, and 91031	91031	[TOE}
June 1	Peterborough 150, trial run from Newcastle to Bounds Green depot, the train formation was 82231, 11275, 11274, 10323, 11273, 12204, and 91031	91031	[TOE}
June 2	Peterborough 150 event, Newcastle to Peterborough, then continuing Kings Cross, the train formation was 82231, 11275, 11274, 10323, 11273, 12204, and 91031	91031	[TOE}
June 3	Peterborough 150 event, test equipment removal.	-	-
June 14	Class 325 EMU, Officer In Charge support for another British Rail test team, Preston - Carnforth	-	-
June 15	Class 325 EMU, Officer In Charge support for another British Rail test team, Preston - Carlisle	-	-
June 17	Class 310 braking tests - preparations at Bletchley depot 310102(3 Car Unit Tests)	-	-
June 18	Class 310 braking tests - 310102 - Test runs - Bletchley - Stafford - Crewe(3 Car Unit Tests)	-	-
June 19	Class 310 braking tests - 310102 - Test runs - Bletchley - Stafford - Crewe(3 Car Unit Tests)	-	-
June 24	Class 310 braking tests - 310102 - (Loaded condition) Test runs - Bletchley - Stafford - Crewe(3 Car Unit Tests)	-	-
June 26	Class 310 braking tests - 310102 - (Loaded condition) Test runs - Bletchley - Stafford - Crewe(3 Car Unit Tests)	-	-
June 30	Class 325 EMU, Officer In Charge support for another British Rail test team, Preston - Carlisle	-	-
July 5	Slip/brake Test - Crewe-Winsford [details not available]	-	[TC2]

Diary1995	Description	Loco	Test Car
July 10 - 11	Nightstock Generator Van (6371) static torsional stiffness ($\Delta Q/Q$) and bogie rotation tests - RTC	-	-
July 21 - 22	Nightstock Generator Van (6371) static sway tests - RTC	-	-
July 24	Nightstock Generator Van (6371) static sway tests (deflated suspension condition) - RTC	-	-
July 30	Nightstock Generator Van (6371) Slip/brake test, Crewe-Winsford	Class 47	[TC2]
Aug 1 - 2	Nightstock Generator Van (6371)static brake tests with the Class 37/6 locomotives – RTC	37601+37602	-
Aug 9	Nightstock Generator Van (6371) dynamic brake tests, Derby-Leicester-Derby (3 round trips) with the Class 37/6 locomotives	37601+37602	-
Aug 16	Nightstock Generator Van (6371) dynamic ride & brake tests, Derby-Doncaster with the Class 37/6 locomotives	37601+37602	-
Aug 17	Nightstock Generator Van (6371) dynamic ride & brake tests, Doncaster-York (3 trips) with the Class 37/6 locomotives	37601+37602	-
Aug 18	Nightstock Generator Van (6371) & Class 37/6 locomotives, test equipment removal at Doncaster	37601+37602	-
Aug 24	Powell Duffryn Supercube Wagon, ride / stress tests - Derby-Cricklewood-Derby, depart 07:15, return 15:30	47981	[TC1]
Sept 6	Buxton Lime Industries BLI19200 JHA Bogie Limestone Hopper Wagon- static torsional stiffness ($\Delta Q/Q$) test– RTC	-	-
Sept 8	Buxton Lime Industries BLI19200 JHA Bogie Limestone Hopper Wagon- static Bogie Rotation test - RTC	-	-
Sept 11	Transfesa 2-axle Wagon - Tare ride test, Derby-Warrington-Derby	Class 47	[TC1]
Sept 14	Transfesa 2-axle Wagon - Laden ride test, Derby-Warrington-Derby	Class 47	[TC1]
Sep 20	Preparations for Class 59 locomotives ride investigations at Merehead	-	-
Sept 21	Class 59 locomotives ride investigations / comparisons (59004) Merehead Quarry - Westbury - Salisbury (2 round trips), portable data recorders used	59004	-
Sept 22	Class 59 locomotives ride investigations / comparisons (59103) Merehead Quarry - Westbury - Salisbury (2 round trips), 59103 fitted with bogie yaw control dampers, portable data recorders used	59103	-
Sept 23	Class 59 locomotives ride investigations / comparisons (59004 / 59103) Merehead Quarry - Westbury, portable data recorders used	59004 + 59103	-

Diary 1995	Description	Loco	Test Car
Sept 26	Blue Circle Cement Wagons (BCC10682 + BCC10694) Laden condition ride test up to 60mile/h between Derby- Crewe - Warrington - Crewe -Derby	47981	[TC1]
Sept 29	Blue Circle Cement Wagons (BCC10682 + BCC10694) static torsional stiffness (ΔQ/Q) tests - RTC	-	-
Nov 4	Iron Ore Tippler Wagons - ride & stability tests, up to 60mile/h between Derby- Crewe - Warrington - Crewe – Derby	47976	[TC1]
Nov 24	Tamper ride / brake test (self-powered) run cancelled at Melton Mowbray and return to RTC (no records of the reason for the test cancellation)	-	-
Nov 29	Tamper ride / brake test (self-powered) run Derby - Old Dalby – Derby	-	-
Nov 30	Class 142 Parking Brake test at Doncaster, static test in the workshops at Wabtec Rail	-	-
Dec 4	Rautarrukki High-Cube Wagon, Y25 1.8m bogies (for Rover Car Parts) - Slip/brake test up to 75 mile/h on the down slow line between Crewe and Winsford, transit back to Derby via Warrington for run-round	Class 47	[TC2]
Dec 6	Tamper ride / brake test (self-powered) run Derby - Old Dalby – Derby	-	-

15. TESTING DIARY - 1996

Diary 1996	Description	Loco	Test Car
Jan 23	Pandrol Jackson Stoneblower 80200, static torsional stiffness ($\Delta Q/Q$) tests – RTC	-	-
Feb 22	Pandrol Jackson Stoneblower 80200, ride tests (self-powered), Derby - Crewe - Derby	-	-
Mar 19	Plasser Power Wagon (YXA-P) preparation for ride tests	-	-
Mar 21	Plasser Power Wagon (YXA-P) ride test between Derby- Crewe - Warrington - Crewe –Derby	Class 47	[TC1]
Mar 25	Plasser Power Wagon (YXA-P) ride test between Derby- Crewe - Warrington - Crewe –Derby	Class 47	[TC1]
Mar 27	Manchester Metrolink tram vehicles - ride investigations, in Manchester - testing on all routes	-	-
Mar 28	Manchester Metrolink tram vehicles - ride investigations, in Manchester - testing on all routes	-	-
Apr 1996	From April to November 1996 I was on a secondment to the Testing Development Group to undertake test data analysis systems development and data analysis systems support for the test teams, I however occasionally supported on-track tests		
May 2	Prototype 102 tonne Hopper Wagon fitted with LTF25 Bogies (JHA 17951) - Slip/brake Test - Crewe-Winsford	Class 47 & Class 86	[TC2]
May 16	Rautaruukki Highcube Wagon, Y25 1.8m bogies (for Rover Car Parts) - ride test, Derby - Warrington	Class 47	[TC1]
May 28	Plasser Power Wagon (YXA-P) - Slip/brake test up to 75 mile/h on the down slow line between Crewe and Winsford, transit back to Derby via Warrington for run-round	Class 47	[TC2]
Nov 15	Eurospine Wagon prototype - Slip/brake test up to 75 mile/h on the down slow line between Crewe and Winsford, transit back to Derby via Warrington for run-round	Class 47	[TC2]
Dec 6	Transfesa Wagon ride test (Dousset type suspension)	Class 47	[TC1]
Dec 16	HOBC (High Output Ballast Cleaner) DR76101 - repeat static torsional stiffness ($\Delta Q/Q$) tests after modifications – RTC	-	-
Dec 17	HOBC (High Output Ballast Cleaner) DR76101 - static bogie rotation tests after modifications – RTC	-	-

16.   TESTING DIARY - 1997

In the latter stages of British Rail privatisation, during February 1997 the former DM&EE testing section business known by this time as Railtest was bought by Serco PLC.

Diary 1997	Description	Loco	Test Car
Feb 2	Mk II and Mk III coach vehicle vibration acceleration data collection to support KE verification - Derby - London Euston – Glasgow	-	-
Feb 3	Mk II and Mk III coach vehicle vibration acceleration data collection to support KE verification - Glasgow - London	-	-
Feb 4	Mk II and Mk III coach vehicle vibration acceleration data collection to support KE verification - London - Manchester – London	-	-
Mar 17	Unilog 60' Container Flat Wagon - ride test - Derby - Warrington - Derby departed at 08:00, returned to RTC at 15:30	Class 47	[TC1]
Apr 1997	From April to June 1997, I again supported the Testing Development Group to undertake test data analysis systems development work.		
July 17	Unilog 60' Container Flat Wagon - ride re-test - Derby - Warrington - Derby	Class 47	[TC1]
Aug 30	Kershaw High Output Ballast Cleaner (HOBC), 76101 testing at Old Dalby test line	-	-
Sept 1	Kershaw High Output Ballast Cleaner (HOBC), 76101 testing at Old Dalby test line	-	-
Sept 4	Plasser 08-RT ride test self-powered - Derby-Warrington-Derby, departed RTC at 11:18 and returned at 18:30.	-	-
Oct 6	Nuclear Flask Wagon (33 70 9985003-0) - static torsional stiffness ($\Delta Q/Q$) test	-	-
Oct 13	Transfesa Wagon with modified suspension, ride test - Derby - Uttoxeter the Transfesa vehicle had a TOPS registration problem that was not found until the test train had left Derby, unfortunately the problem meant that the Transfesa vehicle had to be uncoupled from the Test Car at Uttoxeter and left in the siding there until the registration problem had been resolved. The locomotive and test car returned to Derby	Class 47	[TC1]
Oct 20	Geismar Rail Cleaner - ride / brake testing run Derby-Crewe	-	-
Oct 21	Geismar Rail Cleaner - ride / brake testing run Derby-Crewe	-	-
Oct 24	Tiphook Wagon (Tare) ride test - Derby-Crewe-Carnforth-Derby	Class 47	[TC1]

Diary 1997	Description	Loco	Test Car
Oct 29	Tiphook Wagon (Part Laden) ride test - Derby-Crewe-Carnforth-Derby	Class 47	[TC1]
Nov 6	EWS Wagons (BAA900172 + BBA910567) with 3 - Piece bogies - ride test, Laden condition, Derby - Wigan – Derby	Class 47	[TC1]
Nov 7	EWS Wagons (BAA900172 + BBA910567), Wedge Tests (EDU Workshop)	-	-
Nov 12	EWS Wagons (BAA900172 + BBA910567) 3 - Piece bogies - ride test, Tare condition, Derby - Warrington – Derby	Class 47	[TC1]
Nov 13	EWS Wagons (BAA900172 + BBA910567), Wedge Tests (EDU Workshop)	-	-
Nov 14	EWS Wagons (BAA900172 + BBA910567) with 3 - Piece bogies - ride test, Tare condition, Derby - Warrington – Derby	Class 47	[TC1]
Nov 17	EWS Wagons (BAA900172 + BBA910567), Wedge Tests (EDU Workshop)	-	-
Nov 20	EWS Wagons (BAA900172 + BBA910567) 3 - Piece bogies - ride test, Laden condition, Derby - Warrington – Derby	Class 47	[TC1]
Nov 21	EWS Wagons (BAA900172 + BBA910567), Wedge Tests (EDU Workshop)	-	-
Nov 27	Rolls Royce Nuclear Flask Wagon (33 70 9985003-0), ride test including underframe and secondary bolster stress measurements - Derby - Warrington - Derby, Laden condition	Class 47	[TC6]
Dec 2	EWS BAA Wagon (900172) with 3 Piece bogies, Laden condition (Cobra Brake Blocks, 10" brake cylinders) - Slip/brake test up to 60 mile/h on the down slow line between Crewe and Winsford, transit back to Derby via Warrington for run-round	Class 47	[TC2]
Dec 5	Transfesa 2-axle Wagon ride test - Derby - Warrington – Derby	Class 47	[TC6]
Dec 10	EWS BAA Wagon (900172) with 3 Piece bogies, Laden condition (BBA Friction Brake Blocks, 10" brake cylinders) - Slip/brake test up to 60 mile/h on the down slow line between Crewe and Winsford, transit back to Derby via Warrington for run-round	47976	[TC2]
Dec 11	Transfesa 2-axle Wagon ride test - Derby - Warrington – Derby	Class 47	[TC6]
Dec 17	FPA 2-axle Container Flat Wagon 400201 (Tare) - ride test - Derby - Warrington - Derby	Class 47	[TC6]
Dec 18	FPA 2-axle Container Flat Wagon 400201 (Tare + Container)- ride test - Derby - Warrington - Derby	Class 47	[TC6]

17.   TESTING DIARY - 1998

Diary 1998	Description	Loco	Test Car
Jan 20	Steel carrier (British Steel Strip Products) BSSP4095 ride test, Derby - Sheffield - Derby, this wagon type was originally a Ravenscraig Iron Ore tippler wagon, which was rebuilt by Marcroft Engineering at Stoke	Class 47	[TC6]
Feb 4	EWS BAA Wagon (900172) with 3 Piece bogies, Laden condition (Cobra Brake Blocks, 12" brake cylinders) - Slip/brake test up to 60 mile/h on the down slow line between Crewe and Winsford, transit back to Derby via Warrington for run-round	Class 47	[TC2]
Feb 9	EWS BAA Wagon (900172) with 3 Piece bogies, Tare condition (Cobra Brake Blocks, 12" brake cylinders) - Slip/brake test up to 75 mile/h on the down slow line between Crewe and Winsford, transit back to Derby via Warrington for run-round	Class 47	[TC2]
Feb 10-11	Rolls Royce Nuclear Flask Wagon (33 70 9985003-0) - suspension inspection and friction damping investigation	-	-
Feb 12	Eurospine Wagon (PIGY97001) ride test - Derby-Carnforth - Derby	Class 47	[TC6]
Feb 15	Slip/brake Test - Down Main line - Crewe- Winsford, 110 mile/h maximum, Mk III DVT vehicle	Class 47 & Class 86	[TC2]
Feb 19	Rolls Royce Nuclear Flask Wagon (33 70 9985003-0), ride test - Derby - Warrington - Derby, Tare condition	Class 47	[TC6]
Mar 3	EWS Wagons (BAA900172 + BBA910567) 3 - Piece bogies - ride test, Tare condition, Derby - Carnforth - Derby	Class 47	[TC6]
Mar 7	HOBC (High Output Ballast Cleaner) DR76101 - static bogie rotation tests after further modifications (cab / end bogies only) – RTC	-	-
Mar 9	EWS Wagons (BAA900172 + BBA910567) 3 - Piece bogies - ride test, Laden condition, Derby - Warrington - Derby, delayed testing due to traincrew unavailability	Class 47	[TC6]
Mar 16	EWS Wagons (BAA900172 + BBA910567) 3 - Piece bogies - ride test, Tare (+ 5 tonne) condition, Derby - Carnforth - Derby	Class 47	[TC6]
Mar 20	EWS Wagon BBA910567 - 3 piece bogie tests, Tare condition, Derby - Carnforth - Derby	Class 47	[TC6]
Mar 21	EWS BBA910567 - 3 piece bogie tests, static parking brake test	-	-

Diary 1998	Description	Loco	Test Car
Apr 2	Salmon Wagons (ASF Bogies) - Tare condition ride test - Derby-Warrington-Derby	Class 47	[TC6]
Apr 8	Rolls Royce Nuclear Flask Wagon (33 70 9985003-0) - ride test - Derby - Warrington - Derby, Tare + 5 tonnes (Part Laden) condition	Class 47	[TC6]
Apr 17	EWS BBA Wagon - 3 piece bogie tests (trial with group 5 type spring option), Derby - Carnforth - Derby	Class 47	[TC6]
Apr 23	Salmon Wagons (ASF Bogies) Laden condition ride test - Derby-Warrington-Derby	Class 47	[TC6]
May 21	EWS Wagon BAA900172, 3 - Piece bogies - ride test, Tare condition, Derby - Carnforth - Derby	Class 47	[TC6]
May 28	EWS Wagon BAA900172, 3 - Piece bogies - ride test, Laden condition, Derby - Warrington - Derby	Class 47	[TC6]
June 9	Slip/ Brake Test - Salmon Wagons - cancelled due to locomotive failure at Toton	-	-
June 16	Slip/ Brake Test - Salmon Wagons - test up to 60 mile/h on the down slow line between Crewe and Winsford, transit back to Derby via Warrington for run-round	Class 47	[TC2]
June 29 - 30	EWS Coil Carrier Wagon BYA966001, static tests Laden condition bogie rotation and static side push tests, suspension investigations– RTC	-	-
July 2	EWS Steel Coil Carrier Wagon BYA966001, Laden condition ride test - Derby - Warrington - Derby, late departure due to late locomotive arrival from Toton	37245	[TC6)
July 6 – 8	EWS Coil Carrier Wagon BYA966001, static tests Tare condition torsional stiffness ($\Delta Q/Q$), bogie rotation and static side push tests, suspension investigations– RTC	-	-
July 9	EWS Steel Coil Carrier Wagon BYA966001, Tare condition ride test, Derby - Carnforth - Derby	Class 47	[TC6]
Aug 6	EWS Wagon BAA900172, 3 - Piece bogies, ride test (after 5000 miles in service), Tare condition, Derby - Carnforth - Derby	Class 47	[TC6]
Aug 17	EWS Steel Coil Carrier Wagon BYA966001, Tare condition ride test, Derby - Carnforth - Derby	Class 47	[TC6]
Sept 8	EWS Wagon (Cobra Brake Blocks, Slip/brake test up to 75 mile/h on the down slow line between Crewe and Winsford, transit back to Derby via Warrington for run-round	Class 47	[TC2]
Oct 5	CAIB 2-axle PGA Hopper Wagon Slip/brake test up to 60 mile/h on the down slow line between Crewe and Winsford, transit back to Derby via Warrington for run-round	47976	[TC2]
Oct 15	PD TF25 Bogie Wagon ride test - Derby - Carnforth – Derby	Class 47	[TC6)

Diary 1998	Description	Loco	Test Car
Oct 17	PD TF25 Bogie Wagon ride test - Derby - Carnforth – Derby	Class 47	[TC6)
Nov 12	Salmon Wagons (ASF Bogies with modified friction damping) Tare condition ride test - Derby-Warrington-Derby	Class 47	[TC6)
Nov 17	PD TF25 Bogie wagon laden condition Slip/brake test up to 60 mile/h on the down slow line between Crewe and Winsford, transit back to Derby via Warrington for run-round.	Class 47	[TC2]
Nov 19	Salmon Wagons (ASF Bogies with modified friction damping) Laden condition ride test - Derby-Warrington-Derby	Class 47	[TC6]
Nov 29	Y33 bogie wagon - ride test - Derby-Carnforth-Derby	Class 47	[TC6]
Dec 8	EWS BAA Wagon (900172) Slip/brake test up to 60 mile/h on the down slow line between Crewe and Winsford, transit back to Derby via Warrington for run-round.	Class 47	(TC2)
Dec 11	PD TF25 Bogie wagon tare condition Slip/brake test up to 75 mile/h on the down slow line between Crewe and Winsford, transit back to Derby via Warrington for run-round.	Class 47	[TC2]
Dec 15	Y33 bogie wagon - Slip/brake test up to 60 mile/h on the down slow line between Crewe and Winsford, transit back to Derby via Warrington for run-round.	Class 47	[TC2]

18. TESTING DIARY - 1999

Diary 1999	Description	Loco	Test Car
Jan 7	Class 37 Locomotive - static torsional stiffness ($\Delta Q/Q$) and Bogie Rotation Testing	37421	-
Jan 12	AMEC Tramm - Ride Test		-
Jan 20	Multi-Purpose Vehicle (MPV) DR98901+DR98951, built in Germany by Windhoff, initial testing on Midland Main Line	-	-
Jan 22	Multi-Purpose Vehicle (MPV) DR98901+DR98951, built in Germany by Windhoff, ride and performance testing / transit movement to the Old Dalby test line		-
Jan 25	Multi-Purpose Vehicle (MPV) Self Powered Testing at Old Dalby		-
Feb 1	BAA Wagon Ride Test, tare condition, Derby - Crewe - Carnforth - Crewe - Derby	Class 47	[TC6]
Feb 2	BAA Wagon Slip/brake test, laden condition up to 60 mile/h on the down slow line between Crewe and Winsford, transit back to Derby via Warrington for run-round	Class 47	[TC2]
Feb 5	BAA Wagon Ride Test, laden condition, Derby - Crewe - Carnforth - Crewe - Derby	Class 47	[TC6]
Feb 15	Thrall 2 (FAA) Y33 1.8m bogies - 609001 Container Flat Wagon, Y33 1.8m bogies - torsional stiffness ($\Delta Q/Q$) test	-	-
Feb 16	BAA Wagon fitted with Cobra Brake Blocks - Slip/brake test up to 75 mile/h on the down slow line between Crewe and Winsford, transit back to Derby via Warrington for run-round	Class 47	[TC2]
Feb 26	Thrall 2 (FAA) Y33 1.8m bogies - 609001 Container Flat Wagon, Y33 1.8m bogies - Ride Test	Class 47	[TC6]
Mar 2	KAA Bogie 'Mega 3' Inter-modal Wagon (BCC11501) Slip/brake test up to 75 mile/h on the down slow line between Crewe and Winsford, transit back to Derby via Warrington for run-round	Class 47	[TC2]
Mar 4	Thrall 2 (FAA) Y33 1.8m bogies - 609001 Container Flat - Ride Test - Derby - Crewe - Carnforth	Class 47	[TC6]
Mar 9	KAA Bogie 'Mega 3' Inter-modal Wagon (BCC11501) Slip/brake test up to 75 mile/h on the down slow line between Crewe and Winsford, transit back to Derby via Warrington for run-round	Class 47	[TC2]
Mar 12	Thrall 2 (FAA) Y33 1.8m bogies - 609001 Container Flat Wagon - Ride Test - Derby - Crewe - Carnforth	Class 47	[TC6]

Diary 1999	Description	Loco	Test Car
Mar 23	Thrall 2 (FAA) Y33 1.8m bogies - 609001 Container Flat Wagon Slip/brake test up to 75 mile/h on the down slow line between Crewe and Winsford, transit back to Derby via Warrington for run-round	Class 47	[TC2]
Mar 30	BAA (900172) - Wagon Slip/brake test up to 60 mile/h on the down slow line between Crewe and Winsford, transit back to Derby via Warrington for run-round	Class 47	[TC2]
Apr 1	Thrall 2 (FAA) Y33 1.8m bogies - 609001 Container Flat Wagon, Y33 1.8m bogies - torsional stiffness (ΔQ/Q) investigations test	-	-
Apr 17	Thrall 2 (FAA) Y33 1.8m bogies - 609001 Container Flat Wagon, Y33 1.8m bogies - bogie rotation investigations test	-	-
Apr 13	Thrall 2 (FAA) Y33 1.8m bogies - 609002 / 609053 Container Flat Wagons, Y33 1.8m bogies - torsional stiffness (ΔQ/Q) test	-	-
Apr 14	Thrall 2 (FAA) Y33 1.8m bogies - 609002 / 609053 Container Flat Wagons, Y33 1.8m bogies - bogie rotation tests	-	-
pr 15 - 16	Thrall 2 (FAA) Y33 1.8m bogies - 609002 / 609053 Container Flat Wagons, Y33 1.8m bogies - torsional stiffness (ΔQ/Q) and bogie rotation tests	-	-
Apr 23	Thrall 3 (MBA) 102 tonne Box Wagon 500001 (Swing Motion 1.829m bogies) - Laden Ride Test, Derby - Crewe - Carnforth - Crewe - Derby, locomotive 47565 failure during the test run which caused delay awaiting a rescue locomotive	47565	[TC6]
Apr 30	Thrall 3 (MBA) 102 tonne Box Wagon 500001 (Swing Motion 1.829m bogies) - Ride Test, Derby - Crewe - Carnforth - Crewe - Derby	Class 47	[TC6]
May 14	Geismar VMT 860 PL/UM DR98306+DR98305 - Ride Test - Derby - Crewe - Winsford - Derby [Self Powered], testing up to 60 mile/h.	-	-
July 13	National Power JMA Coal Wagon NP19601- Slip/brake test up to 75 mile/h on the down slow line between Crewe and Winsford, transit back to Derby via Warrington for run-round.	Class 47	[TC2]
July 26 - 27	79 tonne Sleeper Carrying Wagon, Y25C SM 1.8m bogies - Static tests - Delta Q/Q and Bogie Rotation Tests - RTC		-
July 29	KAA Bogie 'Mega 3' Inter-modal Wagon (BCC11501) - Ride Test - Derby - Crewe - Carnforth	Class 47	[TC6]
Aug 4 - 5	79 tonne Sleeper Carrying Wagon, Y25C SM 1.8m bogies - Static tests - Delta Q/Q and Bogie Rotation Tests - RTC		-
Aug 6	79 tonne Sleeper Carrying Wagon, Y25C SM 1.8m bogies - Ride Test - Derby-Crewe-Carnforth	Class 47	[TC6]

Diary 1999	Description	Loco	Test Car
Aug 13	Class 37 static sanding system tests, weighing, bogie rotation testing and ride assessment, Derby to Crewe	37421	–
Aug 19	79 tonne Sleeper Carrying Wagon, Y25C SM 1.8m bogies - Ride Test - Derby-Crewe-Carnforth	Class 47	[TC6]
Sept 1	NISKEY Wagon static torsional stiffness (ΔQ/Q) and bogie rotation tests - RTC		–
Sept 10	NISKEY Wagon - Ride Test - Derby - Warrington - Derby	Class 47	[TC6]
Sept 30	Thrall 2 (FAA) Y33 1.8m bogies - 609001 Container Flat Wagon - Ride Test - Derby-Crewe-Carnforth	Class 47	[TC6]
Oct 5	Thrall 3 (MBA) - 102t Box Wagon 500001 (Swing Motion 1.829m bogies) - Slip/brake test up to 60 mile/h on the down slow line between Crewe and Winsford, transit back to Derby via Warrington for run-round	Class 47	[TC2]
Nov 11	Thrall 3 (MBA) 102t Box Wagon 500001 (Swing Motion 1.829m bogies) - Slip/brake test up to 60 mile/h on the down slow line between Crewe and Winsford, transit back to Derby via Warrington for run-round	Class 47	[TC2]
Nov 30	NACO Hopper Wagon - Ride Test - Derby-Crewe-Carnforth	Class 47	[TC6]
Dec 7	NACO Hopper Wagon - Ride Test - Derby-Crewe-Carnforth	Class 47	[TC6]
Dec 17	Harsco P811 Track Renewal Machine, GPS Type Bogies, DR78901, Ride Test, Derby - Crewe - Carnforth - Crewe - Derby	Class 47	[TC6]
Dec 21	Harsco P811 Track Renewal Machine, GPS Type Bogies, DR78901, Slip/brake test up to 60 mile/h on the down slow line between Crewe and Winsford, transit back to Derby via Warrington for run-round	47798	[TC2]

INDEX

Lightning Source UK Ltd.
Milton Keynes UK
UKHW051615191119
353594UK00006B/47/P